Stefan Rist

Light Propagation in Ultracold Atomic Gases

Stefan Rist

Light Propagation in Ultracold Atomic Gases

Revealing atomic correlations via photon counting

Südwestdeutscher Verlag für Hochschulschriften

Imprint
Any brand names and product names mentioned in this book are subject to trademark, brand or patent protection and are trademarks or registered trademarks of their respective holders. The use of brand names, product names, common names, trade names, product descriptions etc. even without a particular marking in this work is in no way to be construed to mean that such names may be regarded as unrestricted in respect of trademark and brand protection legislation and could thus be used by anyone.

Publisher:
Südwestdeutscher Verlag für Hochschulschriften
is a trademark of
Dodo Books Indian Ocean Ltd., member of the OmniScriptum S.R.L Publishing group
str. A.Russo 15, of. 61, Chisinau-2068, Republic of Moldova Europe
Printed at: see last page
ISBN: 978-3-8381-2563-3

Zugl. / Approved by: Ulm, Universität, Diss. 2011; Barcelona, Universitat Autònoma, Diss. 2011

Copyright © Stefan Rist
Copyright © 2011 Dodo Books Indian Ocean Ltd., member of the OmniScriptum S.R.L Publishing group

Light Propagation in Ultracold Atomic Gases

November 28, 2011

Contents

Introduction   1

1 Quantum Field Theory of Atoms Interacting with Photons   5
    1.1 Classical Hamiltonian 6
    1.2 Quantum Theory 7
        1.2.1 Atomic collisions 9
        1.2.2 Hamiltonian in the length gauge 11
        1.2.3 Second quantization 14
    1.3 Discussion 16

2 Ultracold Atoms in Optical Lattices   19
    2.1 Optical lattices 20
    2.2 Single particle dynamics 22
    2.3 Bose-Hubbard Hamiltonian 26
        2.3.1 Mean-Field Treatment 29
        2.3.2 Bogoliubov Expansion 33
        2.3.3 Particle-Hole Expansion 36
        2.3.4 Random phase approximation 38

3 Photonic Band Structure of a Bichromatic Optical Lattice   47
    3.1 Theoretical model 48
        3.1.1 Hamiltonian 49
        3.1.2 Weak excitation regime 50
        3.1.3 Spin waves 51
    3.2 Monochromatic optical lattice 52
    3.3 Biperiodic optical lattice 55
    3.4 Bichromatic lattice inside a single-mode cavity 58
    3.5 Discussion 62

4 Light Scattering from Ultracold Atoms in an Optical Lattice   65
    4.1 The Model 66
    4.2 Light scattering 69

		4.2.1 Scattering cross section as a function of the atomic state	70
		4.2.2 Numerical results	72
	4.3	Discussion	79
5	**Atomic homodyning**		**81**
	5.1	The Model	83
	5.2	Scattered light intensity	84
		5.2.1 Background contribution	88
		5.2.2 Interference contribution	91
	5.3	A related experiment	94
	5.4	Measuring the atomic field	101
		5.4.1 Temperature Measurement of a Bose-Einstein condensate	105
		5.4.2 Measurement of the superfluid order parameter	108
	5.5	Discussion	112
6	**Summary and Outlook**		**115**
	Appendix		**117**
A	**Derivation of Eq. (5.15)**		**119**
B	**Thermodynamics of trapped Bose-Einstein condensates**		**121**
	B.1	Non interacting case	123
	B.2	Interacting case	124
C	**Derivation of Eq. (5.56)**		**127**
	Bibliography		**131**

Introduction

The wave nature of ultracold matter is perhaps most spectacularly visible in atomic physics experiments, where atomic gases can be prepared at ultralow temperatures, such that the De Broglie wave length becomes of the order of the interparticle distance. This regime where the quantum statistics of the atoms becomes relevant has been first reached experimentally with the achievement of Bose-Einstein condensation in weakly interacting atomic gases in 1995 [1, 2, 3]. These experiments can be considered as benchmark, starting the field of ultracold atoms.

In the subsequent years the main focus of the research of ultracold atomic gases was on exploring the properties of atomic Bose-Einstein condensates. Prominent examples are the interference of two condensates [4], the measurement of long-range phase coherence [5] and the observation of quantized vortices [6, 7, 8] in a Bose-Einstein condensate. We remark here that there are close optical analogues for these experiments, such as the observation of interference of light from two independent sources [9, 10] or the occurrence of optical vortices in the modal structure of some lasers [11].

Drawing on these studies, increasing attention has been devoted to atomic gases with strong correlations, in the regime in which the many-body dynamics essentially determines the system properties. This has been achieved for instance with optical lattices. It was shown theoretically by Jaksch *et al.* in 1998 [12] that the lowest lying excitations of ultracold bosonic atoms in a dispersive optical lattice can be described by the Bose-Hubbard Hamiltonian. The implementation of the Bose-Hubbard Hamiltonian in an optical lattice and the observation of the quantum phase transition from superfluid to Mott insulator phase has been first reported by Greiner *et al.* in 2002 [13]. It opened the door to study ultracold atomic gases in a regime that is typically studied in solid state and condensed matter physics and constitutes a benchmark in the research with ultracold atomic gases. Further reading about the progress so far in the research of ultracold atomic gases can be found in [14, 15] and references therein.

A crucial point for comparing experiment with theoretical predictions is the characterization of the many-body state of the atoms. For ultracold atoms detection of the quantum state is often performed by releasing the atoms from the confining region and measuring their density distribution after the time of flight to the detecting region by absorption imaging. For sufficiently long time of flight the measured density distribution is proportional to the initial momentum distribution in the trap [14]. Beyond measuring the initial momentum distribution several experimental techniques allow one to extract distinct physical quantities characterising the quantum state of the atoms. Applying a Bragg-pulse initially leads to a well defined momentum and energy transfer to the atoms and permits the study of collective excitations of the system [16, 17, 18, 19], thereby measuring the structure form factor, which is essentially the Fourier transform of the density-density correlation function of the gas [20, 21, 22]. The single-particle correlation function of the atomic gas can be determined by spatially resolved outcoupling of atoms from the trap and measuring the resulting interference pattern [5]. This technique has been used to study the emergence of off diagonal long range order in Bose-Einstein condensation [23]. It was also shown theoretically that one may be able to measure the superfluid fraction of an ultracold atomic gas by applying a light induced vector potential to simulate rotation [24].

Further measurement techniques based on the so called noise correlation measurements take into account quantum fluctuations by comparing absorption images of different experimental runs [25]. Applying noise correlation measurements to the absorption image of two interfering identical atomic samples it has been shown that one can completely determine the quantum state of the atomic system for one and two dimensional Bose-liquids [26, 27, 28]. In general they have proven to be a valuable tool for extracting additional information about the many-body state of the atoms [29, 30, 31].

Despite the vast amount of information that can be obtained about the atomic systems using absorption imaging after time of flight, the main drawback is its destructive character. An alternative approach to determine the quantum state of the ultracold atomic gas is the measurement of elastically scattered light from the atomic system. The motivation for this route to determine the quantum state of ultracold atomic gases is manifold. Several theoretical works pointed out that the observation of photons scattered by ultracold atoms may provide complementary information on the quantum state of the atoms compared to the usual time of flight measurements [32, 33, 34, 35, 36, 37, 38, 39]. For dispersive light-matter interaction such measurements may be non-destructive for the quantum state of the atoms [35, 40, 41, 42]. Hence they may set the basis for feedback schemes [41, 43], which could be used to engineer the quantum state of the atoms.

Using optical detection schemes it is difficult to find a setup which gives information about the single-particle correlation function of the atomic gas. This is due to the fact that for dispersive light-matter interaction the first order coherence of the scattered light contains information about the density-density correlations of the atoms [44]. Indeed optical detection of the single-particle correlation function of an atomic system has only been considered for Fermions so far [45, 46]. It is an open question how one may obtain this information for a gas of bosons by measuring the scattered light. Also no proposal exists which may allow for measuring the single particle correlation function of an ultracold atomic gas in a non-destructive way, independent on the quantum statistics of the atoms.

In the present thesis we study ultracold bosonic gases interacting dispersively with the electromagnetic field, and analyse the kind of information about the many-body state of the atoms that may be obtained by measuring the scattered photons. The general setup we consider is the following. A weak probe field drives the atoms such that its effect on the atomic evolution can be neglected. The scattered light is then calculated as a function of various parameters such as scattering angle, frequency or intensity. Using variations of this setup we show how to measure the density-density correlation function and the excitation spectrum of the atomic gas, as well as the mean value of the atomic field operator by detecting the emitted photons. We use ultracold atoms in optical lattices as model system for our investigations and shown how the presented schemes may allow one to monitor the quantum phase transition of an ultracold bosonic gas in an optical lattice from superfluid to Mott-insulator phase. We note that their application is not restricted to ultracold atoms in optical lattices and extensions of them may allow for the measurement of the single-particle correlation function of an arbitrary system of ultracold bosonic atoms.

The thesis is organized as follows. In chapter 1 we review the basic theory for the light-matter interaction which forms the basis for our investigations of light scattering from ensembles of ultracold atoms. The theoretical model is based on second quantization, thus taking into account the quantum statistics of the atoms.

In chapter 2 we review the dynamics of ultracold atoms in optical lattices which form one of the basic systems for our investigations. We first introduce some basic properties of the physics of a single particle in a periodic potential. Considering the case of many atoms it is shown that the ground state and the lowest lying excitations of ultracold bosonic atoms confined by a dispersive optical lattice can be described by the Bose-Hubbard Hamiltonian [12]. We then report some theoretical techniques to perturbatively calculate the excitation spectrum and the eigenstates of the Bose-Hubbard Hamiltonian, which will be used in the following chapters.

In chapter 3 we study the photonic band structure of a chain of fixed pointlike atoms in a biperiodic configuration interacting with the electromagnetic vacuum. We calculate the photonic spectra and the probe transmission if the atoms are in free space and inside a standing wave optical resonator. We study how the photonic spectra are modified as a function of the interparticle distance. It is shown how one can get information about the atomic configuration by measuring the transmitted light signal.

In chapter 4 we investigate the spectrum of the scattered light from ultracold atoms in an optical lattice illuminated by a weak laser. We calculate the photonic scattering cross section as a function of energy and direction of emission along the Mott-insulator superfluid phase transition. Thereby we take the finite tunneling rate of the atoms and the effect of photon recoil into account when evaluating the scattering cross section. The interference between the finite atomic tunneling rate and the photon induced hopping is visible in the heights of the Bragg peaks and we show that this effect is measurable in the superfluid phase. We compare our analytical results using the techniques presented in Chapter 2 with numerical results where we diagonalize exactly the Bose-Hubbard Hamiltonian for a small number of atoms and wells. It is shown that frequency resolved measurement of the scattered light at different scattering angles may reveal the spectrum of the atomic system and hence the many-body state of the atoms.

In chapter 5 we show how the mean value of the atomic field operator $\langle\psi(\mathbf{r},t)\rangle$ of an ultracold atomic system may be determined by means of photo detection. For this purpose we consider a gas of ultracold atoms trapped at two spatially separated regions in space. Using two Raman lasers to couple atoms out of the systems we show that it is possible to measure the mean value of the atomic field operator $\langle\psi(\mathbf{r},t)\rangle$ of the two systems by measuring the scattered light intensity in one of the lasers modes. We compare our general theory to the experimental results of Saba *et al.* [47] and find agreement between experimental data and our theoretical results. It is shown that measuring the scattered light is equivalent to directly measuring the atoms by absorption imaging in this setup. We then argue how an extension of this setup might be used to measure the temperature of a Bose-Einstein condensate or to monitor the superfluid to Mott-insulator phase transition for a gas of ultracold atoms in an optical lattice.

In Chapter 6 we summarize the main results and give an outlook of the work presented in this thesis. We relate current directions of research to our results and discuss possible extensions of the presented studies.

Chapter 1

Quantum Field Theory of Atoms Interacting with Photons

In this chapter we review the quantum mechanical description for the interaction of an atomic gas with the electromagnetic field. When the atoms are at sufficiently high temperatures and low densities, the atom-photon interaction reported in textbooks [48, 49, 50] provides an excellent description of the dynamics, which has been confirmed by experiments [49, 51]. In this case the atoms interact individually with the photons. Collective dynamics may emerge e.g. if the interparticle distance is much smaller than the resonant wavelength of the atomic dipole transition which couples to the electromagnetic field. In this regime spontaneous phase-locking of the atomic dipoles leads to superradiant behaviour of the atomic sample [52]. However the collective emission of light from the atomic sample leading to superradiance is established via the electromagnetic field and not due to an intrinsic collective behaviour of the atoms.

Experiments in atomic physics have reached regimes where the temperature of an atomic gas is sufficiently low, such that the thermal de Broglie wavelength of the particles becomes comparable with the interatomic distance. In this regime the quantum statistics of the gas is crucial and has to be taken into account also for describing the light matter interaction [14].

In this thesis we will focus on the photons-atoms dynamics at ultralow temperatures, when the many-body effects and quantum statistics of the atoms are relevant. The theory to employ should hence account for the many-body effects in the interactions between atoms and light, and in particular their quantum statistics.

A theory that describes the interaction of a gas of quantum degenerate atoms with the quantized electromagnetic field is most conveniently cast into second quantized form. In this case the quantum statistics of the atoms will be automatically taken into account using the proper commutation relation for the atomic field operators [53]. Since it forms the basis for the rest of this thesis we illustrate in this chapter the main steps that lead to a second quantized description for the interaction of an ultracold atomic gas with the electromagnetic field, as has been first reviewed in 1994 by Lewenstein et al. [54].

We start with the classical Hamiltonian for N charged pointlike particles in an electromagnetic field in Coulomb gauge. Quantization of the electromagnetic field and replacement of the coordinates of the particles and their canonical momenta with their respective operators [55] leads to the basic Hamiltonian for a gas of atoms interacting with the electromagnetic field in the velocity gauge. Applying the dipole and rotating wave approximation we transform the Hamiltonian to the length gauge by means of a Goeppert-Mayer transformation [56]. We end the chapter by making some critical remarks about the presented theory and its limitations.

1.1 Classical Hamiltonian

The classical Hamiltonian of N charged pointlike particles with masses m_j and charges e_j at positions $\mathbf{r}_j$ interacting with the electromagnetic field in Coulomb gauge ($\nabla \cdot \mathbf{A} = 0$) is given by [50]

$$H = \sum_j \frac{1}{2m_j}\left(\mathbf{p_j} - \frac{e_j}{c}\mathbf{A_j}\right)^2 + \frac{1}{2}\sum_{j\neq k}\frac{e_j e_k}{|\mathbf{r}_j - \mathbf{r}_k|} + \frac{1}{8\pi}\int d\mathbf{r}(\mathbf{E}^{\perp 2} + \mathbf{B}^2), \quad (1.1)$$

where $\mathbf{A}_j$ is the vector potential taken at the position of the jth particle $\mathbf{r}_j$ and

$$\mathbf{B} = \nabla \times \mathbf{A}, \quad (1.2a)$$

$$\mathbf{E}^{\perp} = -\frac{1}{c}\frac{\partial \mathbf{A}}{\partial t}, \quad (1.2b)$$

are the magnetic and transverse electric field respectively. The three terms on the right hand side of Eq. (1.1) describe the kinetic energy of the atoms moving in an electromagnetic field, the static Coulomb interaction between the charged particles and the energy of the electromagnetic field in vacuum respectively.

The integral in Eq. (1.1) runs over the whole volume of the system which we assume to be a cube of length L. Expanding the vector potential in plane waves, assuming periodic boundary conditions, one finds [48]

$$\mathbf{A}(\mathbf{r},t) = \sum_\lambda A_\lambda \left[\alpha_\lambda(t)e^{i\mathbf{k}_\lambda \cdot \mathbf{r}} + \alpha_\lambda^*(t)e^{-i\mathbf{k}_\lambda \cdot \mathbf{r}}\right]\hat{\mathbf{e}}_\lambda, \quad (1.3)$$

where the sum over λ has to be seen as a sum over all discrete wave vectors $k_{x,y,z} = 2\pi n_{x,y,z}/L$, with $n_{x,y,z}$ being some integer, and polarizations $\hat{\mathbf{e}}_\lambda \perp \mathbf{k}_\lambda$. In Eq. (1.3) we expanded the vector potential in modes with linear polarizations such that $\hat{\mathbf{e}}_\lambda$ and the amplitudes A_λ real valued. Inserting Eq. (1.3) in Eq. (1.2) one finds from Maxwells equations for the electric and magnetic field the following equations of motion for the normal variables α_λ in the free field case

$$\dot{\alpha}_\lambda(t) = -i\omega_\lambda \alpha_\lambda(t), \quad (1.4)$$

with the dispersion relation $\omega_\lambda = |\mathbf{k}_\lambda| c$.

Now we will discuss the charges. The particles we consider in this thesis are bosonic alkali-metal atoms, thus neutrally charged and with a single valence electron and integer spin. In this case the atoms can be modelled to consist of heavy nuclei (completely filled shells) with mass m and effective charge $-e$ at positions $\mathbf{R}_j$ and a much lighter single valence electron of mass m_e and charge e at positions $\mathbf{r}_j$, moving in the potential of the core[1]. If the density of the gas is not too high, such that the distance between distinct atoms is always much larger than the extension of the atoms, $\langle |\mathbf{R}_j - \mathbf{r}_j| \rangle \ll \langle |\mathbf{R}_k - \mathbf{R}_l| \rangle \ \forall \{j, k \neq l\}$ [2], each atom can be considered as a dipole with moment $\mathbf{d}_j = |e|(\mathbf{r}_j - \mathbf{R}_j)$. Hence the leading order terms in the Coulomb interaction in Eq. (1.1) are well approximated by the screened Coulomb potential between the valence electrons and their cores and the dipole part of the Coulomb potential between two different atoms. All the other terms can be included by an effective short range atom-atom potential $V_{jm}(\mathbf{R}_j, \mathbf{R}_m)$. Thus the classical Hamiltonian of a gas of N hydrogen like atoms in the single electron approximation in an electromagnetic field is given by

$$H = \sum_j \left(\frac{(\mathbf{P}_j + \frac{e}{c}\mathbf{A}(\mathbf{R}_j))^2}{2m} + \frac{(\mathbf{p}_j - \frac{e}{c}\mathbf{A}(\mathbf{r}_j))^2}{2m_e} + V_{ec}(\mathbf{R}_j, \mathbf{r}_j) \right) + \frac{1}{2}\sum_{j \neq k} V_{jk} + H_{dip} + H_{emf}, \quad (1.5)$$

where $\mathbf{p}_j$ and $\mathbf{P}_j$ are the canonical momenta for the valence electrons and the cores respectively. The term $V_{ec}(\mathbf{R}_j, \mathbf{r}_j)$ describes the screened Coulomb potential between the valence electron and core of the jth atom and

$$H_{dip} = \frac{1}{2}\sum_{j \neq k} \left(\frac{\mathbf{d}_j \cdot \mathbf{d}_k - 3(\hat{\mathbf{n}}_{jk} \cdot \mathbf{d}_j)(\hat{\mathbf{n}}_{jk} \cdot \mathbf{d}_k)}{|\mathbf{R}_j - \mathbf{R}_k|^3} + \frac{4\pi}{3}\mathbf{d}_j \cdot \mathbf{d}_k \delta(\mathbf{R}_j - \mathbf{R}_k) \right), \quad (1.6)$$

$$H_{emf} = \frac{1}{8\pi}\int d\mathbf{r}(\mathbf{E}^{\perp 2} + \mathbf{B}^2), \quad (1.7)$$

give the static dipole-dipole interaction [57] and the energy of the electromagnetic field in vacuum, respectively.

1.2 Quantum Theory

In order to obtain a quantum mechanical description from the classical Hamiltonian Eq. (1.5) one has to replace the canonical conjugate variables of the electrons and cores with their respective operators and the Poisson brackets with commutators [55, 58].

For a better description of typical experimental setups we will also include an external confining potential $V_t(\mathbf{R}_j, \mathbf{r}_j, \alpha)$ for the atoms. Since the valence electrons are

---
[1] Note that e is the charge of the electron and is negative.
[2] Here $\langle O \rangle$ means the classical time average of O

subject to the strong Coulomb potential of the core they will not feel the external confinement, such that the trap potential will only affect the center of mass motion of the atoms, which is essentially the motion of the cores. The dependence on $\mathbf{r}_j$ only reflects the fact that the trap potential might be dependent on the internal state of the atoms and on the spin of the atom α [59]. We will not consider the case of velocity dependent external potentials.

The quantization of the electromagnetic field follows the rules of second quantization in vacuum. Using Eq. (1.2), where the vector potential is given by Eq. (1.3), in Eq. (1.7) one finds that the Hamiltonian of the electromagnetic field is given by a sum of harmonic oscillators, each determined by its wave vector and polarization. One proceeds with the quantization by replacing

$$\alpha_\lambda(t) \to a_\lambda(t),$$
$$\alpha_\lambda^*(t) \to a_\lambda^\dagger(t), \qquad (1.8)$$

where the operators $a_\lambda^\dagger$ and a_λ are creation and annihilation operators for a photon in the mode at frequency ω_λ, wave vector $\mathbf{k}_\lambda$ and polarization $\hat{\mathbf{e}}_\lambda \perp \mathbf{k}_\lambda$. They obey the commutation relation $[a_\lambda, a_{\lambda'}^\dagger] = \delta_{\lambda,\lambda'}$. Choosing the amplitudes $A_\lambda = \sqrt{\frac{2\pi\hbar c^2}{\omega_\lambda L^3}}$ in Eq. (1.3) the corresponding Hamiltonian for the free electromagnetic field in vacuum in quantum theory then reads

$$H_{emf} = \sum_\lambda \hbar\omega_\lambda a_\lambda^\dagger a_\lambda, \qquad (1.9)$$

where we neglected the zero point contribution of each mode[3]. With the replacement Eq. (1.8) one finds for the quantized transverse vector potential[4]

$$\mathbf{A}(\mathbf{r},t) = \sum_\lambda \sqrt{\frac{2\pi\hbar c^2}{\omega_\lambda \mathcal{V}}} \left[a_\lambda(t) e^{i\mathbf{k}_\lambda \cdot \mathbf{r}} + a_\lambda^\dagger(t) e^{-i\mathbf{k}_\lambda \cdot \mathbf{r}} \right] \hat{\mathbf{e}}_\lambda, \qquad (1.10)$$

where $\mathcal{V} = L^3$ is the quantization volume and the sum has to be read as a sum over all wave vectors and polarizations. The derivation of the Hamiltonian of the free electromagnetic field Eq. (1.9) and the expression for the quantized vector potential Eq. (1.10) can be derived rigorously by starting from a Lagrangian formulation of classical electrodynamics [48].

Since the mass of the valence electron is much smaller than the mass of the core $m_e \ll m$ we can take $\mathbf{R}_j$ to be the coordinate of the center of mass motion and $\mathbf{P}_j$ its conjugate momentum. In this case $\mathbf{p}_j$ essentially corresponds to the conjugate momentum of the relative coordinates $\bar{\mathbf{r}}_j = \mathbf{r}_j - \mathbf{R}_j$. All corrections to these approximations are of order

---

[3]The zero point contribution for each mode is the energy of the ground state of the quantized electromagnetic field. Since it is an additive constant it only produces an overall energy offset and can be discarded as long as the quantization volume is constant [50, 58].

[4]In the Coulomb gauge the vector potential does not have a longitudinal component.

1.2 Quantum Theory

$O(m_e/m)$. With these approximations the Hamiltonian that describes N interacting hydrogen-like atoms in an external confinement reads

$$H_{vg} = \sum_j \left(H_{CM}^j + H_0^j + H_{af}^j \right) + H_{aa} + H_{emf}, \qquad (1.11)$$

where the subscript vg stands for velocity gauge and

$$H_{CM}^j = \frac{\mathbf{P}_j^2}{2m} + V_t(\mathbf{R}_j, \mathbf{r}_j, \alpha), \qquad (1.12)$$

$$H_0^j = \frac{\mathbf{p}_j^2}{2m_e} + V_{ec}(\bar{\mathbf{r}}_j), \qquad (1.13)$$

describe the center of mass motion in the external potential $V_t(\mathbf{R}_j, \mathbf{r}_j, \alpha)$ and the motion of the electron around the core respectively. The term H_{af}^j describes the interaction between the jth[5] atom and the electromagnetic field and reads

$$H_{af}^j = -\frac{e}{m_e c} \mathbf{p}_j \cdot \mathbf{A}(\mathbf{R}_j) + \frac{e^2}{2m_e c^2} \mathbf{A}(\mathbf{R}_j)^2, \qquad (1.14)$$

where we have neglected factors of order $O(m_e/m)$ [54] and made the dipole approximation $\mathbf{A}(\mathbf{r}_j) \approx \mathbf{A}(\mathbf{R}_j)$[6]. The term H_{aa} describes the atom-atom interaction via the static dipole interaction as given in Eq. (1.6) (with the coordinates replaced by their respective operators) and atomic collisions such that

$$H_{aa} = H_{dip} + \frac{1}{2} \sum_{j \neq k} V_{col}(\mathbf{R}_j, \mathbf{R}_k). \qquad (1.15)$$

1.2.1 Atomic collisions

Interactions between the atoms are important for the dynamics and they determine the quantum ground state of the system. At ultralow temperatures collisions between bosonic atoms are dominated by s-wave scattering. In this regime the whole scattering process is determined by a single parameter, the s-wave scattering length [60]. In the following we will provide the main steps that lead to an effective potential that describes s-wave scattering of ultracold bosonic atoms.

The collision of two atoms can be reduced to a single particle problem if one uses centre of mass and relative coordinates. Thus it can be described as a particle with reduced mass moving in an external scattering potential $U(r)$. Taking $U(r)$ to be

---

[5] Here the index j serves as labeling but has no real physical meaning since all the atoms are indistinguishable.

[6] The dipole approximation can be made since the extension of an atom $a_0 \sim \langle |\mathbf{R}_j - \mathbf{r}_j| \rangle$ is much smaller than the resonant wavelength $a_0 \ll \lambda_0$ such that the change of the vector potential within a_0 can be neglected [48].

spherically symmetric the wavefunction of a particle initially moving in the positive z direction has the asymptotic form

$$\psi(r,\theta) \approx e^{ikz} + \frac{f(\theta)}{r} e^{ikr} \qquad (1.16)$$

for large distances $r \gg r_0$, where r_0 is the effective range of $U(r)$. Eq. (1.16) consists of the incoming wave and an outgoing spherical wave, where $f(\theta)$ is the scattering amplitude which determines the strength of the scattering and depends on the scattering angle θ. The scattering cross section is given by

$$\frac{d\sigma}{d\Omega} = |f(\theta)|^2. \qquad (1.17)$$

Since the scattering potential $U(r)$ is spherically symmetric we can write the general solution of the wavefunction as [60]

$$\psi(r,\theta) = \sum_{l=0}^{\infty} A_l P_l(\cos\theta) R_{k,l}(r). \qquad (1.18)$$

Here A_l are constants, $P_l(x)$ are Legendre polynomials and $R_{k,l}(r)$ are the radial wavefunctions. The quantum number l labels the angular momentum of the solution of the wavefunction $\psi(r,\theta)$. If we require that the general solution Eq. (1.18) has the asymptotic form Eq. (1.16) one finds that the scattering amplitude is given by [60]

$$f(\theta) = \frac{1}{2ik} \sum_{l=0}^{\infty} (2l+1)(e^{2i\delta_l} - 1) P_l(\cos\theta), \qquad (1.19)$$

where the coefficients δ_l are the phase shifts of the radial wavefunctions $R_{k,l}(r)$. The phase shifts δ_l are determined by the radial Schrödinger equation for the functions $R_{k,l}(r)$ and in principle depend on the details of the scattering potential $U(r)$. Eq. (1.19) can also be written in the form $f(\theta) = \sum_{l=0}^{\infty} f_l$ where f_l is the contribution for each partial wave with angular quantum number l. For ultracold atoms scattering takes place at very low energies and one finds that the contributions of the partial waves to the scattering amplitude behave as

$$f_l \sim k^{2l}, \qquad (1.20)$$

where the only condition on the scattering potential $U(r)$ is that it falls off sufficiently fast with distance r [60]. We see from Eq. (1.20) that for low energies (small k) the $l = 0$ (s-wave) term dominates over all other contributions. Thus it is sufficient for bosonic atoms [7] to take only the s-wave contribution into account when describing atomic collisions if the temperature for the atoms is sufficiently low. Since Eq. (1.20) is

---

[7] For fermions s-wave scattering is excluded due to the Pauli principle

1.2 Quantum Theory

independent on the details of the scattering potential $U(r)$ one can show that the whole scattering problem is described by a single parameter, the scattering length a_s, which is defined as

$$a_s = -\lim_{k \to 0} f(\theta). \tag{1.21}$$

One can thus replace the true interatomic collision potential $V_{col}(\mathbf{R}_j, \mathbf{R}_k)$ in Eq. (1.15) with an effective zero range potential

$$V_{col}(\mathbf{R}_j, \mathbf{R}_k) \to g\delta(\mathbf{R}_j - \mathbf{R}_k). \tag{1.22}$$

The parameter g can be related to the s-wave scattering length a_s via [61]

$$g = \frac{4\pi\hbar^2 a_s}{m}. \tag{1.23}$$

The s-wave scattering length a_s is usually determined experimentally since for its theoretical calculation one needs the details of the interatomic potential $V_{col}(\mathbf{R}_j, \mathbf{R}_k)$.

1.2.2 Hamiltonian in the length gauge

Hamiltonian Eq. (1.11) has some disadvantages due to the long range static dipole-dipole interaction H_{dip} and the nonlinear $\mathbf{A}^2(\mathbf{R_j})$ terms. These can be removed if one uses the so called length gauge. The Hamiltonian in the length gauge can be obtained by means of the Goeppert Mayer transformation [48] which reads

$$T = \exp\left(\frac{-\imath e}{\hbar c} \sum_j \bar{\mathbf{r}}_j \cdot \mathbf{A}(\mathbf{R}_j)\right). \tag{1.24}$$

The Hamiltonian and wavefunction in the old and new gauge are related to each other via

$$H_{lg} = TH_{vg}T^\dagger, \tag{1.25}$$

$$|\psi\rangle_{lg} = T|\psi\rangle_{vg}. \tag{1.26}$$

To obtain the Hamiltonian in the length gauge we note that

$$T\mathbf{p}_j T^\dagger = \mathbf{p}_j + \frac{e}{c}\mathbf{A}(\mathbf{R}_j), \tag{1.27}$$

$$Ta_\lambda T^\dagger = a_\lambda + \imath e\sqrt{\frac{2\pi}{\hbar\omega_\lambda \mathcal{V}}}\sum_j (\bar{\mathbf{r}}_j \cdot \hat{\mathbf{e}}_\lambda)e^{-\imath \mathbf{k}_\lambda \cdot \mathbf{R}_j}. \tag{1.28}$$

Thus the free electromagnetic field transforms as

$$TH_{emf}T^\dagger = H_{emf} - \sum_j \mathbf{d}_j \cdot \mathbf{E}^\perp(\mathbf{R}_j) + H_d, \tag{1.29}$$

where the last term is given by

$$H_d = \frac{2\pi e^2}{V} \sum_{j,k} \sum_{\lambda} (\mathbf{r}_j \cdot \hat{\mathbf{e}}_\lambda)(\mathbf{r}_k \cdot \hat{\mathbf{e}}_\lambda) e^{i\mathbf{k}_\lambda \cdot \mathbf{R}_{jk}}, \quad (1.30)$$

with $\mathbf{R}_{jk} = \mathbf{R}_j - \mathbf{R}_k$. Now we will show that H_d exactly cancels out the long range static dipole term in H_{dip} given in Eq.(1.6). For this purpose we perform the sum over the polarization and transform the sum into an integral in the standard way to get

$$H_d = 2\pi e^2 \sum_{j,n} \frac{1}{8\pi^3} \int d\mathbf{k} \left(\bar{\mathbf{r}}_j \cdot \bar{\mathbf{r}}_n - \frac{(\bar{\mathbf{r}}_j \cdot \mathbf{k})(\bar{\mathbf{r}}_n \cdot \mathbf{k})}{\mathbf{k}^2} \right) e^{i\mathbf{k} \cdot \mathbf{R}_{jn}}. \quad (1.31)$$

Comparison with the transverse delta function [48]

$$\delta_{ij}^\perp(\mathbf{r}) = \frac{1}{8\pi^3} \int d\mathbf{k} \left(\delta_{ij} - \frac{k_i k_j}{\mathbf{k}^2} \right) e^{i\mathbf{k} \cdot \mathbf{r}} \quad (1.32)$$

$$= \frac{2}{3}\delta_{ij}\delta(\mathbf{r}) - \frac{1}{4\pi r^3}\left(\delta_{ij} - \frac{3r_i r_j}{r^2}\right) \quad (1.33)$$

shows that

$$H_d = 2\pi \sum_{j \neq k} (\mathbf{d}_j \cdot \mathbf{d}_k) \, \delta(\mathbf{R}_j - \mathbf{R}_k) - H_{dip}. \quad (1.34)$$

We discarded the terms $j = m$ in the sum since they correspond to infinite dipole self-energy terms. Thus we see that the static dipole terms cancel out in the new gauge and the remaining contact term can be included into the short range collision terms. The Hamiltonian in the length gauge has the same form as Eq. (1.11) with the difference that now the $\mathbf{p}_j$ are kinetic momenta proportional to the velocity of the electron and

$$H_{af} = -\sum_j \mathbf{d}_j \cdot \mathbf{E}^\perp(\mathbf{R}_j), \quad (1.35)$$

$$H_{aa} = \frac{1}{2} \sum_{j \neq k} \left(g(\mathbf{r}_j, \mathbf{r}_k) + 4\pi \mathbf{d}_j \cdot \mathbf{d}_k \right) \delta(\mathbf{R}_j - \mathbf{R}_k). \quad (1.36)$$

The dependence of the strength of atomic collisions $g(\mathbf{r}_j, \mathbf{r}_k)$ on the coordinates of the electrons indicates that the scattering length depends on the internal state of the atoms. We note that, apart from collisions, the atoms only interact via the exchange of transverse photons. Due to the resonant character of the matter light interaction in a typical experimental setup it is convenient to restrict the Hilbert space to a manifold of electronic ground and excited states which are eigenstates of the Hamiltonian $H_0^j = \frac{\mathbf{p}_j}{2m} + V_{ec}(\bar{\mathbf{r}}_j)$. We label these states by $|g, \alpha\rangle_j$ and $|e, \beta\rangle_j$, where α and β are internal

1.2 Quantum Theory

spin indices. Introducing the following atomic raising and lowering operators

$$\sigma_{ge}^{\alpha\beta}(j) = |g,\alpha\rangle_j\langle e,\beta|_j, \tag{1.37}$$

$$\sigma_{eg}^{\beta\alpha}(j) = \sigma_{ge}^{\alpha,\beta}(j)^\dagger, \tag{1.38}$$

$$\sigma_{gg}^{\alpha\alpha'}(j) = |g,\alpha\rangle_j\langle g,\alpha'|_j, \tag{1.39}$$

$$\sigma_{ee}^{\beta\beta'}(j) = |e,\beta\rangle_j\langle e,\beta'|_j, \tag{1.40}$$

we rewrite the total Hamiltonian as

$$H = H_0 + H_{af} + H_{aa} + H_{emf}, \tag{1.41}$$

with

$$H_0 = \sum_j \left(\frac{\mathbf{P}_j}{2m} + \sum_\beta [\omega_0 + V_{te}(\mathbf{R}_j, \beta)] \sigma_{ee}^{\beta\beta}(j) + \sum_\alpha V_{tg}(\mathbf{R}_j, \alpha) \sigma_{gg}^{\alpha\alpha(j)} \right). \tag{1.42}$$

Here ω_0 is the transition frequency between the ground state manifold and the excited state manifold. For simplicity we assume ω_0 to be independent on spin indices. The trap potential for the ground and excited states are labelled tg and te respectively. Using the rotating wave approximation the atom field interaction is given by

$$H_{af} = \sum_j \sum_{\alpha,\beta} \sum_\lambda \hbar C_\lambda^{\beta\alpha} \sigma_{eg}^{\beta\alpha}(j) a_\lambda e^{i\mathbf{k}_\lambda \cdot \mathbf{R}_j} + \text{H.c.}, \tag{1.43}$$

with

$$C_\lambda^{\beta\alpha} = \sqrt{\frac{2\pi\omega_\lambda}{\hbar\mathcal{V}}} \left(\mathbf{D}^{\beta\alpha} \cdot \hat{\mathbf{e}}_\lambda \right), \tag{1.44}$$

where $\mathbf{D}^{\beta\alpha} = -\imath\langle e,\beta|\mathbf{d}_j|g,\alpha\rangle$ is the dipole moment between the ground state $|g,\alpha\rangle$ and the excited state $|e,\beta\rangle$. $\mathbf{D}^{\beta\alpha}$ is independent of j and can be taken to be real [48]. The atomic collision part can be written as

$$H_{aa} = H_{gg} + H_{eg}^{col} + H_{eg}^{dip} + H_{ee}, \tag{1.45}$$

where

$$H_{gg} = \frac{g_{gg}}{2} \sum_{j\neq k} \sum_{\alpha,\alpha'} \delta(\mathbf{R}_j - \mathbf{R}_k) \sigma_{gg}^{\alpha\alpha}(j) \sigma_{gg}^{\alpha'\alpha'}(k) \tag{1.46}$$

describes collisions between atoms in the ground state manifold. The term

$$H_{ee} = \frac{g_{ee}}{2} \sum_{j\neq k} \sum_{\beta,\beta'} \delta(\mathbf{R}_j - \mathbf{R}_k) \sigma_{ee}^{\beta\beta}(j) \sigma_{ee}^{\beta'\beta'}(k) \tag{1.47}$$

gives the collisions between atoms in the excited state manifold, while

$$H_{eg}^{col} = \frac{1}{2} \sum_{j \neq k} \sum_{\alpha,\beta} \delta(\mathbf{R}_j - \mathbf{R}_k) \left(g_{eg}^{(1)} \sigma_{eg}^{\beta\alpha}(j) \sigma_{ge}^{\alpha\beta}(k) + g_{eg}^{(2)} \sigma_{ee}^{\beta\beta}(j) \sigma_{gg}^{\alpha\alpha}(k) \right) + \text{H.c.} \quad (1.48)$$

describes collisions between atoms in the ground-and atoms in the excited state manifold and

$$H_{eg}^{dip} = \frac{1}{2} \sum_{j \neq k} \sum_{\alpha,\beta} 4\pi \delta(\mathbf{R}_j - \mathbf{R}_k) \left| \mathbf{D}^{\beta\alpha} \right|^2 \sigma_{eg}^{\beta\alpha}(j) \sigma_{ge}^{\alpha\beta}(k) + \text{H.c} \quad (1.49)$$

is responsible for the dipolar interaction between the atoms. We take the scattering length to be spin independent and also discard the possibility that the atoms change their internal spin state during the collision.

If the quantum statistics of the atoms does not play a role, Hamiltonian Eq. (1.41) is well suited to describe the interaction of an atomic gas with the quantized electromagnetic field. We will use this description in Chapter 3 where we take the atoms to be tightly confined at fixed positions in space such that one can neglect their motional degrees of freedom. If the motion of the atoms cannot be neglected and their thermal de Broglie wavelength is comparable with the interparticle distance, as it is the case in most problems studied in this thesis, the quantum statistics of the atoms is important and a description of the light matter interaction in second quantized form is more convenient.

1.2.3 Second quantization

In order to describe the light matter interaction in second quantized form we introduce the field operators $\psi_g(\mathbf{r}, \alpha)$ ($\psi_e(\mathbf{r}, \beta)$) and $\psi_g(\mathbf{r}, \alpha)^\dagger$ ($\psi_e(\mathbf{r}, \beta)^\dagger$) which describe the annihilation and creation of an atom at position $\mathbf{r}$ in the internal state $|g, \alpha\rangle$ ($|e, \beta\rangle$). For bosonic atoms the atomic field operators fulfill the commutation relations

$$\left[\psi_i(\mathbf{r}, \alpha), \psi_j^\dagger(\mathbf{r}', \alpha') \right] = \delta_{ij} \delta_{\alpha\alpha'} \delta(\mathbf{r} - \mathbf{r}'), \quad (1.50a)$$

$$\left[\psi_i(\mathbf{r}, \alpha), \psi_j(\mathbf{r}', \alpha') \right] = \left[\psi_i^\dagger(\mathbf{r}, \alpha), \psi_j^\dagger(\mathbf{r}', \alpha') \right] = 0, \quad (1.50b)$$

where $i, j = (e, g)$. For fermionic atoms the commutators in Eq. (1.50) have to be replaced with anticommutators. Performing the second quantization of Hamiltonian Eq. (1.41) is a standard procedure in quantum field theory [53]. The Hamiltonian in the length gauge in second quantized form for N atoms interacting with the electromagnetic field reads [54]

$$H = H_g + H_e + H_{af} + H_{gg} + H_{ee} + H_{eg} + H_{emf}, \quad (1.51)$$

1.2 Quantum Theory

with

$$H_g = \sum_\alpha \int d\mathbf{r}\, \psi_g^\dagger(\mathbf{r},\alpha) \left(\frac{-\hbar^2 \nabla^2}{2m} + V_{tg}(\mathbf{r},\alpha) \right) \psi_g(\mathbf{r},\alpha), \quad (1.52\text{a})$$

$$H_e = \sum_\beta \int d\mathbf{r}\, \psi_e^\dagger(\mathbf{r},\beta) \left(\frac{-\hbar^2 \nabla^2}{2m} + \hbar\omega_0 + V_{te}(\mathbf{r},\beta) \right) \psi_e(\mathbf{r},\beta), \quad (1.52\text{b})$$

where H_g (H_e) determines the unperturbed atomic dynamics of the atoms in the internal state $|g,\alpha\rangle$ ($|e,\beta\rangle$) moving in the trapping potential $V_{tg}(\mathbf{r},\alpha)$ ($V_{te}(\mathbf{r},\beta)$). The Hamiltonian

$$H_{af} = \sum_{\lambda,\alpha} \hbar C_\lambda^{\beta\alpha} \int d\mathbf{r}\, \psi_e^\dagger(\mathbf{r},\beta) \psi_g(\mathbf{r},\alpha) a_\lambda e^{i\mathbf{k}_\lambda \cdot \mathbf{r}} + \text{H.c.} \quad (1.52\text{c})$$

describes the interaction of the atoms with the electromagnetic field in the rotating wave and dipole approximations and

$$H_{emf} = \sum_\lambda \hbar \omega_\lambda a_\lambda^\dagger a_\lambda \quad (1.52\text{d})$$

is the energy of the free electromagnetic field, where we discarded the constant energy shift due to the quantum vacuum. The terms

$$H_{gg} = \frac{g_{gg}}{2} \sum_{\alpha\alpha'} \int d\mathbf{r}\, \psi_g^\dagger(\mathbf{r},\alpha) \psi_g^\dagger(\mathbf{r},\alpha') \psi_g(\mathbf{r},\alpha') \psi_g(\mathbf{r},\alpha), \quad (1.52\text{e})$$

$$H_{ee} = \frac{g_{ee}}{2} \sum_{\beta,\beta'} \int d\mathbf{r}\, \psi_e^\dagger(\mathbf{r},\beta) \psi_e^\dagger(\mathbf{r},\beta') \psi_e(\mathbf{r},\beta') \psi_e(\mathbf{r},\beta), \quad (1.52\text{f})$$

$$H_{eg} = \sum_{\alpha,\beta} g_{eg}^{\alpha\beta} \int d\mathbf{r}\, \psi_e^\dagger(\mathbf{r},\beta) \psi_g^\dagger(\mathbf{r},\alpha) \psi_g(\mathbf{r},\alpha) \psi_e(\mathbf{r},\beta), \quad (1.52\text{g})$$

are responsible for s-wave scattering between atoms in internal states $|i\rangle$ and $|j\rangle$, where $g_{eg}^{\alpha\beta} = g_{eg}^{(1)} + g_{eg}^{(2)} + 4\pi|\mathbf{D}^{\beta\alpha}|^2$. In the rest of this thesis we will restrict to atoms where the relevant levels for the light matter interaction can be described either by a two level system, in which case the spin indices α,β can be omitted, or a lambda type of system as indicated in Fig. (1.1). In the latter case the annihilation operator for atoms in state $|j\rangle$ will be labeled $\psi_j(\mathbf{r})$ with $j = (1,2,e)$ and the spin indices will be omitted. The coupling strength of the dipolar transitions between ground state $|j\rangle$ and excited state $|e\rangle$ is given by

$$C_\lambda^{e,j} = \sqrt{\frac{2\pi\omega_\lambda}{\hbar\mathcal{V}}} \left(\mathbf{D}^{e,j} \cdot \hat{\mathbf{e}}_\lambda \right), \quad (1.53)$$

where $\mathbf{D}^{e,j}$ is the dipole moment between ground state $|j\rangle$ and the excited state $|e\rangle$.

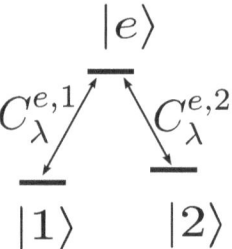

Figure 1.1: Sketch of a lambda-type level system with two groundstates $|1\rangle$, $|2\rangle$ and one excited state $|e\rangle$. The dipolar transition between a ground state $|j\rangle$ and the excited state $|e\rangle$ via the absorption of photon with frequency ω_λ and polarization $\hat{\mathbf{e}}_\lambda$ has coupling strength $C_\lambda^{e,j}$.

1.3 Discussion

In deriving Eq.(1.51) several approximations have been made which are usually well justified in the single atom case. Some of them however are questionable in the many-body case where atoms might get close to each other. One of these approximations is the two level approximation. While for a single atom this approximation is well justified it surely breaks down if two atoms collide since in this case their electronic wavefunctions become deformed and mixed. The mixing of the electronic wavefunctions is most spectacularly visible in the formation of ultracold molecules [62].

Another crucial assumption is the fact that the quantum statistics of the atoms is simply taken to be determined by the total spin of the atoms. Nevertheless, even bosonic atoms are composed by fermionic particles and are therefore composite bosons. The Pauli exclusion principle which holds for the fermionic components of the atoms is completely neglected in such a treatment as well as exchange processes, where two bosonic atoms interchange some of their fermionic components.

These problems have already been mentioned in [54] and it was suggested that one should try to find a better effective potential for atomic collisions that includes such effects. Recently a theory has been developed by M. Combescot [63] that is capable to describe accurately the many-body physics for composite bosons. Within this theory it is shown that exchange processes are not energy-like processes and cannot be described in the usual Hamiltonian way. Hence a theory describing the physics of ultracold atomic gases that takes into account for the effects of exchange interaction in atomic collisions should probably be constructed along the lines indicated in [63]. This may open up a way to describe in a clean way atomic collisions taking into account the exchange of electrons. This would be especially desirable for ensembles of ultracold Rydberg atoms [64] where the outer valence electron is weakly bound such that exchange processes might

1.3 Discussion

become relevant and Hamiltonian Eq. (1.51) might not give an accurate description of the system.

In the rest of this thesis we will work with Hamiltonian Eq. (1.51) and assume that it yields a good description for the physical systems we will discuss. This is especially the case for low densities and atoms for which the valence electron is strongly bound to the atomic core.

Chapter 2

Ultracold Atoms in Optical Lattices

A major advance in the research with ultracold atomic gases has been made by means of optical lattices [14]. Optical lattices are periodic potentials, which arise from the mechanical effect of light-atom interaction. They were first realized in 1992, when atoms were trapped by an optical standing wave. Light scattering from the system confirmed the spatial long range order of the atoms in the periodic potential [65, 66].

Experiments with atoms in optical lattices focus on several directions, ranging from cooling to the simulation of the dynamics so far predicted in condensed matter systems [14, 15]. One direction, which has attracted several experimental and theoretical efforts is the simulation of the Bose-Hubbard model [67]. This research was motivated by the observation that the lowest lying excitations of ultracold bosonic atoms in a dispersive optical lattice can be described by the Bose-Hubbard Hamiltonian [12, 13].

The Bose-Hubbard Hamiltonian plays a prominent role in condensed matter theory and is used, e.g., to describe Cooper pairs of electrons undergoing Josephson tunneling between superconducting islands or helium atoms on a substrate [68]. It exhibits a quantum phase transition from a superfluid to a Mott-insulator phase at zero temperature, $T = 0$. In the case of ultracold atoms in optical lattices the superfluid state is characterized by a Poisson distributed number of atoms at each site and a well defined phase relation between the atomic wavefunctions in different lattice sites. In the Mott-insulator phase each lattice site has a well defined and equal number of particles and there is an energy gap between the ground state and the first excitations, leading to vanishing compressibility [68].

In this thesis we will discuss ultracold bosonic atoms in optical lattices as model system which is dispersively coupled to light in order to measure atomic correlations by photodetection. For this purpose we will review some important theoretical tools to describe the ground state and the dynamics of ultracold bosonic atoms in optical lattices.

The present chapter is organized as follows: After shortly discussing how optical lattices are realized experimentally we review some properties of a single particle moving in a periodic potential. We then consider the case of an ultracold bosonic gas in a dispersive optical lattice and show that it can be described by the Bose-Hubbard Hamiltonian. We review some general properties of the Bose-Hubbard Hamiltonian and introduce various techniques for studying the ground state and dynamics, which will be instrumental for the studies in the rest of this thesis.

2.1 Optical lattices

The conservative potential of optical lattices originates from the mechanical effects of light on atoms. More precisely, the potential is a position dependent dynamical Stark shift of the electronic ground state of atoms whose dipolar transition couples with the standing wave of a far detuned laser. We model now this interaction drawing on the description of light-matter interaction introduced in chapter1.

We assume that two laser beams with frequency ω_L, wave vector $\pm k_L \hat{e}_x$ and equal polarization couple to an atomic dipole transition with ground state $|g, \alpha\rangle$, excited state $|e, \beta\rangle$ and transition frequency ω_0 as depicted in Fig. (2.1). The laser fields are described by coherent states of the corresponding electromagnetic field mode with amplitude α_L, such that the mean number of photons is given by $|\alpha_L|^2$. The light matter interaction is given in Eq. (1.52c), where the coupling constant of the dipolar transition driven by the laser is denoted by $C_L^{\beta\alpha}$ and is defined in Eq. (1.44). In the regime in which the atom-laser coupling is sufficiently weak, corresponding to the condition $|C_L^{\beta\alpha} \alpha_L| \ll |\omega_0 - \omega_L|$, one can adiabatically eliminate the excited state from the equations of motion of the ground state in second-order perturbation theory in the small parameter $|C_L^{\beta\alpha} \alpha_L|/|\omega_0 - \omega_L|$ [69]. The dynamics of the atoms in the electronic ground state $|g, \alpha\rangle$ are then described by the effective Hamiltonian

$$H_{\text{eff}} = H_g + H_{gg} + H_{emf} + H'_{\text{int}}. \tag{2.1}$$

where H_g, H_{gg} and H_{emf} are given in Eq. (1.52a), Eq. (1.52e) and Eq. (1.52d) respectively. The interaction term Eq. (1.52c) takes the form

$$H'_{\text{int}} = \hbar \sum_{\lambda,\lambda'} \frac{C_\lambda^{\beta\alpha} C_{\lambda'}^{\beta\alpha'}}{\omega_{\lambda'} - \omega_0} a_\lambda^\dagger a_{\lambda'} \int d\mathbf{r} e^{i\mathbf{q}\cdot\mathbf{r}} \psi_g^\dagger(\mathbf{r},\alpha) \psi_g(\mathbf{r},\alpha). \tag{2.2a}$$

Equation (2.2a) describes the absorption of a photon in the mode λ' and wave vector $\mathbf{k}_{\lambda'}$ and the emission into the mode λ and wave vector $\mathbf{k}_\lambda$, with $\mathbf{q} = \mathbf{k}_{\lambda'} - \mathbf{k}_\lambda$. We assume the light matter interaction to be dispersive such that $\gamma \ll |\omega_0 - \omega_L|$, which is here fulfilled as the light is sufficiently far off resonance from the atomic transition. In this case photon emission into other modes than the laser can be neglected and the

2.1 Optical lattices

effective light matter interaction reads

$$H'_{\text{int}} \approx 4\hbar \frac{\Omega_0^2(\alpha,\beta)}{\Delta} \int d\mathbf{r} \psi_g^\dagger(\mathbf{r},\alpha) \sin^2(2k_L x) \psi_g(\mathbf{r},\alpha), \tag{2.2b}$$

where we introduced the Rabi frequency $\Omega_0(\alpha,\beta) = |C_L^{\beta\alpha}|$, the detuning $\Delta = \omega_L - \omega_0$ and we have neglected a constant offset in energy. We note that the light matter

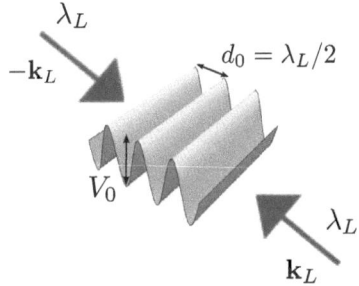

Figure 2.1: Schematic illustration of a one-dimensional optical lattice. Two counter-propagating laser beams with wave vectors $\pm \mathbf{k}_L$, wavelength λ_L and equal polarization couple far off resonantly to an atomic dipolar transition forming a periodic potential for the atoms. The lattice constant of the resulting periodic potential is given by $d_0 = \lambda_L/2$.

interaction described by Eq. (2.2b) is equivalent to an effective external potential for the atoms in state $|g,\alpha\rangle$ of the form

$$V(x) = V_0 \sin^2 \frac{\pi x}{d_0}, \tag{2.3}$$

where $d_0 = \frac{\lambda_L}{2}$ is the lattice spacing with $\lambda_L = \frac{2\pi}{|\mathbf{k}_L|}$ being the wavelength of the laser that creates the optical lattice and $V_0 = 4\hbar \frac{\Omega_0^2(\alpha,\beta)}{\Delta}$ is the lattice depth, which is determined by the strength of the coupling between laser and atomic dipole and its sign is determined by the detuning Δ. In the case of a blue detuned optical lattice $\Delta > 0$ the atoms tend to be confined at the minima of the field intensity. If on the other hand the optical lattice is red detuned $\Delta < 0$ the atoms will be attracted to the antinodes of the standing wave where the light intensity is maximal.

Higher dimensional lattices can be created in a similar way by superposing more laser beams[14]. Using laser beams with different polarizations that couple to different internal states of the atoms for instance, one can create potentials which can manipulate atoms in different ways, depending on the internal quantum number α [70]. Taking these cases into account the effective Hamiltonian can we written as

$$H^a_{eff} = \sum_\alpha \int d\mathbf{r}\psi_g^\dagger(\mathbf{r},\alpha) \left(\frac{-\hbar^2 \nabla^2}{2m} + V_{tg}(\mathbf{r},\alpha) + V(\mathbf{r},\alpha) \right) \psi_g(\mathbf{r},\alpha) + H_{gg}, \qquad (2.4)$$

where $V(\mathbf{r},\alpha)$ is a trapping potential for the atoms which emerges from the interaction with the laser.

2.2 Single particle dynamics

In this section we review some basic properties of the dynamics of a single particle of mass m moving in a periodic potential $V(x+d_0) = V(x)$, such as an atom in an optical lattice, where d_0 is the lattice constant. We consider the system to be confined in a box of length $L = Md_0$ where M is the number of primitive cells and use periodic boundary conditions. The Hamiltonian of the system is given by

$$H = \frac{p^2}{2m} + V(x). \qquad (2.5)$$

The eigenfunctions $\psi_n(x)$ of H obey the time independent Schrödinger equation

$$H\psi_n(x) = E_n \psi_n(x), \qquad (2.6)$$

where E_n is the nth energy eigenvalue. We expand the eigenfunctions $\psi_n(x)$ in a superposition of plane waves

$$\psi_n(x) = \sum_k c_k^n e^{ikx}, \qquad (2.7)$$

where the wave vectors k are determined by the boundary condition and given by $k = \frac{2\pi}{Md_0} j$, with j some integer. From the normalization $\int_0^L dx |\psi_n(x)|^2 = 1$ one finds the condition $\sum_k |c_k^n|^2 = 1$ for the coefficients c_k^n. Since the lattice potential is periodic its Fourier expansion will contain only vectors of the reciprocal lattice which we here denote $G = \frac{2\pi}{d_0} j$, where j is some integer[1]. The potential can hence be written as

$$V(x) = \sum_G V_G e^{iGx}, \qquad (2.8)$$

with

$$V_G = \frac{1}{d_0} \int dx e^{-iGx} V(x). \qquad (2.9)$$

Using Eq. (2.7) and Eq. (2.8) in Eq. (2.6) one obtains

$$\left(\frac{\hbar^2}{2m} k^2 - E \right) c_k^n + \sum_G V_G c_{k-G}^n = 0. \qquad (2.10)$$

---

[1] In three dimensions a wave vector $\mathbf{K}$ is defined to be a reciprocal lattice vector if $e^{i\mathbf{K}\cdot\mathbf{R}} = 1$ for all lattice vectors $\mathbf{R}$ [71].

2.2 Single particle dynamics

Due to the periodicity of the potential only coefficients c_k^n, c_{k-G}^n are coupled that differ by some reciprocal lattice vector G. It is convenient to write $k = q + G'$ where G' is a reciprocal lattice vector which is chosen such that q lies in the interval $[-\frac{\pi}{d_0}, \frac{\pi}{d_0}]$, the first Brillouin zone. Due to the periodic boundary conditions there are exactly M wave vectors q lying in the first Brillouin zone and Eq. (2.10) is given by

$$\left(\frac{\hbar^2}{2m}(q-G')^2 - E_n\right) c_{q-G'}^n + \sum_G V_{G-G'} c_{q-G}^n = 0. \tag{2.11}$$

Equation (2.11) shows that it is possible to separate the original problem into M independent equations, one for each q within the first Brillouin zone. The initial eigenvalue problem can be written in the form

$$H\psi_q^n(x) = E_q^n \psi_q^n(x), \tag{2.12}$$

with the Bloch functions $\psi_q^n(x)$, which can be written as

$$\psi_q^n(x) = \sum_G c_{q+G}^n e^{i q + G x} \tag{2.13}$$

$$= u_n(x) e^{i q x}, \tag{2.14}$$

with the function $u_q^n(x) = \sum_G c_{q+G}^n e^{iGx}$ of the same periodicity as the lattice, $u_q^n(x + d_0) = u_q^n(x)$. Equation (2.14) is the statement of the Bloch theorem [71]. Due to the periodicity of the potential the spectrum shows a band structure where the eigenenergies E_q^n are continuous functions of the quasi-momentum q and separated into distinct bands, labeled by n. In one dimension the energy ranges for different bands are separated by a finite gap [71].

The normalization of the Bloch functions is usually chosen such that

$$\frac{2\pi}{d_0} \int_{-d_0/2}^{d_0/2} \psi_q^n(x) \bar{\psi}_{q'}^m(x) = \delta_{n,m} \delta_{q,q'}, \tag{2.15}$$

where $\delta_{n,m}$ is the Kronecker delta.

Now we calculate the energy spectrum of a single particle subject to the potential Eq. (2.3). In this case Eq. (2.11) reads

$$\left((q+G)^2 + \frac{V}{2}\right) c_{q+G}^n - \frac{V}{4}\left(c_{q+G+G_0}^n + c_{q+G-G_0}^n\right) = \mathcal{E}_q^n c_{q+G}^n, \tag{2.16}$$

with $G_0 = \frac{2\pi}{d_0}$,

$$V = \frac{\pi^2 V_0}{d_0^2 E_R}, \tag{2.17}$$

$$\mathcal{E}_q^n = \frac{\pi^2 E_q^n}{d_0^2 E_R}, \tag{2.18}$$

and E_R is the recoil energy,

$$E_R = \frac{\hbar^2 k_L^2}{2m}. \tag{2.19}$$

Equation (2.16) shows that the determination of coefficients c_{q+G}^n consists in diagonalizing a tridiagonal matrix. Figure (2.2) displays the band structure for two different lattice depth.

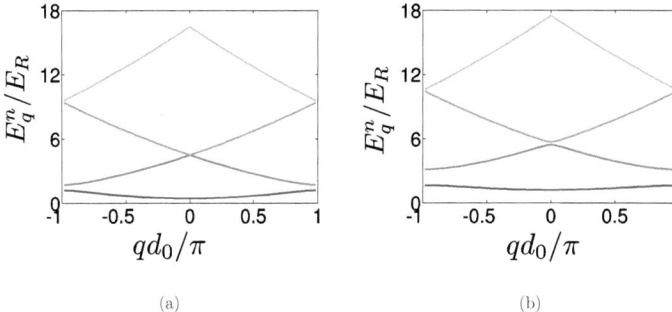

(a) (b)

Figure 2.2: Energy E_q^n in units of E_R as a function of q (in units of π/d_0) in the first Brillouin zone for the potential Eq. (2.3), with a) $V_0 = E_R$ b) $V_0 = 3E_R$. We note that the gap between different bands increases with the lattice depth.

In order to solve the dynamics of an atomic system subject to a periodic potential it can be convenient to introduce the basis of the so called Wannier functions $w_l^n(x)$. The Wannier function $w_l^n(x)$ is centered around $x_l = ld_0$ and describes a particle that is localized in the lth lattice minima in the nth band. The Wannier functions are the Fourier transform of the Bloch functions

$$w_l^n(x) = \left(\frac{2\pi}{d_0}\right)^{1/2} \frac{1}{M} \sum_{q \in BZ} \psi_q^n(x) e^{\imath q x_l}, \tag{2.20}$$

where M is the total number of lattice sites and the sum runs over all vectors q in the first Brillouin zone. The Wannier functions in one dimension can be taken to be real

2.2 Single particle dynamics

and they form a complete orthonormal basis [72]

$$\int dx\, w_l^n(x) w_{l'}^{n'}(x) = \delta_{ll'}\delta_{nn'}, \qquad (2.21)$$

$$\sum_{n,l} w_l^n(x) w_l^n(x') = \delta(x-x'). \qquad (2.22)$$

Hence we can also express the Bloch functions in terms of the Wannier functions

$$\psi_q^n(x) = \left(\frac{d_0}{2\pi}\right)^{1/2} \sum_l w_l^n(x) e^{-iqx_l}, \qquad (2.23)$$

which is the inverse Fourier transform of Eq. (2.20).

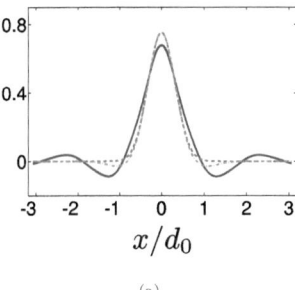

(a)

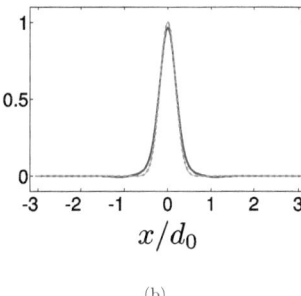

(b)

Figure 2.3: Wannier function defined in Eq. (2.20) and Gaussian approximations defined in Eq. (2.24), Eq. (2.25) as a function of position x in units of the lattice spacing d_0 for the potential Eq. (2.3). The solid blue line corresponds to the numerically evaluated Wannier function Eq. (2.20) in the lowest band, the green dashed line to the ordinary Gaussian approximation Eq. (2.24) and the red dot-dashed line to the modulated Gauss function Eq. (2.25) for $l = 0$. a) $V_0 = E_R$ b) $V_0 = 10 E_R$

For a sufficiently deep lattice one can expand the lattice potential around the minima to second order in x. Approximating the Wannier function in the lowest band $n = 1$ with the ground state of the harmonic potential leads to the usual Gaussian approximation for the Wannier function $w_l^1(x) \approx g_l(x)$ with

$$g_l(x) = \frac{1}{\sqrt{\xi\sqrt{\pi}}} e^{-\frac{(x-ld_0)^2}{2\xi^2}}, \qquad (2.24)$$

where $\xi = \frac{d_0}{\pi}\left(\frac{E_R}{V_0}\right)^{1/4}$ is the oscillator length of the ground state in the harmonic potential. The Gaussian *Ansatz* has to be handled with care since the Wannier functions behave as $e^{-|x|}$ for $x \to \infty$ [72] and thus decay much more slowly than the Gaussian $g_l(x)$. Another disadvantage of the Gaussian approximation is that the wavefunctions for different sites are not orthogonal

$$\int dx g_l(x) g_m(x) = f_o(|l-m|) \neq 0 \text{ for } l \neq m,$$

where $f_o(|l-m|)$ is the overlap between the two Gaussian centered around x_l and x_m. For deep lattices, such that the lattice depth $V_0 \gg E_R$ is of the order of several recoil energies, the overlap becomes very small. In this case the overlap between nearest neighbours is the dominant one, such that $f_o(1) \gg f_o(n) \; \forall \; n > 1$. This gives one the possibility to make the Gaussian approximately orthogonal, defining so called modified Gaussian functions $\tilde{g}_l(x)$ where [73]

$$\tilde{g}_l(x) = g_l(x) - \frac{f_o(1)}{2}(g_{l+1}(x) + g_{l-1}(x)). \qquad (2.25)$$

The overlap between neighbouring modified Gaussian functions $\int dx \tilde{g}_l(x) \tilde{g}_{l+1} = O(f_o(2))$ is of the order of the overlap between next nearest neighbours of the Gaussian approximation, which is negligible for the parameter range where this approximation is applicable. Figure (2.3) displays $g_l(x)$ and $\tilde{g}_l(x)$, which are compared with the Wannier functions calculated numerically. One sees in Fig. (2.3a) that for shallow lattices with lattice depth $V_0 \sim E_R$ the Gaussian *Ansatz* and also the modified Gaussian *Ansatz* fit only poorly the Wannier function, as in this case the mismatch at the tails is relevant. For deep lattices, e.g. $V_0 = 10 E_R$ the Gaussian approximations $g_l(x)$ and $\tilde{g}_l(x)$ are in good agreement with the numerically calculated Wannier function, as visible in Fig. (2.3b).

2.3 Bose-Hubbard Hamiltonian

Ultracold atoms confined by a dispersive optical lattice can be seen in a simplified picture as a system where the atoms are restricted to a discrete spatial grid, where the grid points are the lattice minima. At ultralow temperatures, such that the thermal energy is well below the lattice depth, the atoms can move from site to site only via quantum tunneling, where the tunneling matrix element between two adjacent sites is denoted by J, and interact via a repulsive on-site interaction U as shown in Fig. (2.4).

The Hamiltonian that determines the dynamics of the atoms in the optical lattice in second quantization is obtained from Eq. (2.4) and reads

$$H_g = \int d\mathbf{r} \psi_g^\dagger(\mathbf{r}) \left(\frac{-\hbar^2 \nabla^2}{2m} + V_l(\mathbf{r})\right) \psi_g(\mathbf{r}) + \frac{b_{gg}}{2}\int d\mathbf{r} \psi_g^\dagger(\mathbf{r})\psi_g^\dagger(\mathbf{r})\psi_g(\mathbf{r})\psi_g(\mathbf{r}),$$

2.3 Bose-Hubbard Hamiltonian

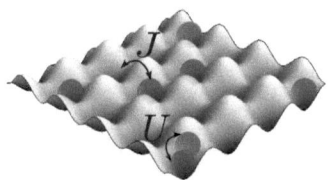

Figure 2.4: Schematic illustration of a gas of ultracold bosonic atoms in a 2D optical lattice as described by the Bose-Hubbard Hamiltonian. Atoms can tunnel from site to site with tunneling strength J and interact via the repulsive on-site interaction U.

where we have only considered a single electronic ground state and have hence omitted the spin index α. For simplicity we restrict to the case of a one-dimensional optical lattice. In experimental setups this can be achieved by using a three-dimensional optical lattice and increase the lattice depth in the radial direction such that dynamics in the transverse plane is practically frozen out [74]. In this case one can approximate the potential for the atoms in the directions of tight confinement by an harmonic potential and we obtain for the lattice potential

$$V_l(\mathbf{r}) = V_0 \sin^2\left(\frac{\pi x}{d_0}\right) + \frac{1}{2} m \omega_r (y^2 + z^2), \qquad (2.26)$$

where ω_r is the frequency of the harmonic trap which tightly confines the transverse motion. We assume that the atomic wavefunctions are well localized at the lattice minima, such that the tight-binding approximation can be applied. Furthermore, at ultralow temperatures and not too strong interparticle interactions, the atomic gas is in the lowest band of the periodic potential and in the ground state of the radial oscillator. In this restricted Hilbert space the atomic field operator can be decomposed as

$$\psi_g(\mathbf{r}) = \phi_0(\rho) \sum_l w_l(x) b_l, \qquad (2.27)$$

where $w_l(x)$ describes an atom in the lowest band centered at position $x_l = l d_0$, which we will refer to as lattice site l, and which is given in Eq. (2.20). We omit the band index n since we restrict to the lowest band. The sum in Eq. (2.27) runs over all lattice sites, and $\phi_0(\rho) = \exp(-\rho^2/2\xi_r^2)/(\xi_r \sqrt{\pi})$ is the ground state wavefunction of the radial oscillator ($\rho = \sqrt{y^2 + z^2}$) with $\xi_r = \sqrt{\hbar/m\omega_r}$. The operator b_l annihilates an atom at site l and fulfills the standard bosonic commutation relations $[b_l, b_{l'}^\dagger] = \delta_{l,l'}$. Using this decomposition in Eq. (2.26) we obtain the Bose-Hubbard Hamiltonian in one dimension

$$H'_g = -J \sum_l b_l^\dagger (b_{l-1} + b_{l+1}) + \frac{U}{2} \sum_l n_l(n_l - 1), \qquad (2.28)$$

where we allowed only for nearest-neighbour hopping and restricted to on-site atom-atom interactions. Here $n_l = b_l^\dagger b_l$ is the atomic number operator at site l. The coefficients J and U denote the hopping term and the on-site interaction strength respectively and are given by

$$J = -\int dx\, w_l(x) \left(-\frac{\hbar^2 \nabla^2}{2m} + V(x)\right) w_{l+1}(x), \qquad (2.29)$$

$$U = b_{gg} \frac{m\omega_r}{4\pi\hbar} \int dx\, w_l(x)^4, \qquad (2.30)$$

with the Wannier functions chosen to be real. Restricting to the ground state for the transverse motion of the atoms, as assumed in Eq. (2.27), is justified by taking $J, U\langle n\rangle \ll \hbar\omega_r$, where $\langle n\rangle = \text{tr}\{b_l^\dagger b_l\}$ is the mean site occupation. The ground state of Hamiltonian Eq. (2.28) exhibits a quantum phase transition from a superfluid state for $J \gg U$ to a Mott-insulator state for $J \ll U$ at a critical ratio J_c/U_c. The qualitative behaviour of the system in the two phases can be understood if one considers the limiting cases of either vanishing on-site interaction and nonzero tunneling strength ($U = 0, J \neq 0$) or finite on-site interaction and vanishing tunneling strength ($J = 0, U \neq 0$). In both cases Hamiltonian Eq. (2.28) can be solved exactly. If the on-site interaction U is zero the ground state is given by

$$|\psi_0^{(0)}\rangle = \frac{1}{\mathcal{N}} \left(\sum_{l=1}^M b_l^\dagger\right)^N |0\rangle, \qquad (2.31)$$

where $\mathcal{N}$ is the normalization, N the total number of atoms, M the total number of lattice sites and $|0\rangle$ is the vacuum. Equation (2.31) describes a state where each atom is delocalized over the whole lattice with equal occupation probability for all sites. In momentum space the superfluid state can be seen as Bose-Einstein condensate, where all atoms occupy the lowest momentum eigenstate of the optical lattice. In general, for $J > J_c$ and $U \neq 0$ the superfluid state is characterized by the on-site atomic occupation which follows a Poisson distribution, such that $\Delta n = \langle n^2\rangle - \langle n\rangle^2 = \langle n\rangle$, and the atomic system has a nonvanishing order parameter $\langle b_l\rangle \neq 0$.

For vanishing tunneling J, Hamiltonian Eq. (2.28) is diagonal in the basis of number states (Fock basis) and the ground state $|\psi_0^{(0)}\rangle$ is given by a product of Fock states with equal occupation $g = N/M$ at each site,

$$|\psi_0^{(0)}\rangle = \prod_{l=1}^M \frac{\left(b_l^\dagger\right)^g}{\sqrt{g!}} |0\rangle = |g, g, ...g, g\rangle, \qquad (2.32)$$

2.3 Bose-Hubbard Hamiltonian

where we have assumed commensurable filling. In general, in the Mott-insulator state number fluctuations and the order parameter of the system vanish $\Delta n = \langle b_l \rangle = 0$. The two behaviours are shown pictorially in Fig. (2.5).

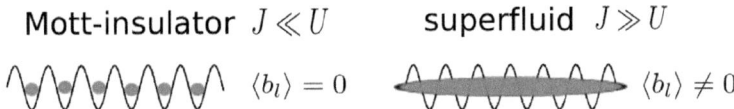

Figure 2.5: Sketch of the density distribution of the ground state of an ultracold atomic gas in an optical lattice in the Mott-insulator and superfluid phase. The Mott insulator state is a product of Fock states with well defined atom number at each lattice site, such that $\langle b_l \rangle = 0$. In the superfluid state the atoms are delocalized over the whole lattice with well defined phase relation between different lattice sites one has a non vanishing order parameter $\langle b_l \rangle \neq 0$.

In the following we will introduce some basic theoretical methods to investigate the stationary and dynamical properties predicted by the Bose-Hubbard Hamiltonian in d dimensions. The generalization of Hamiltonian Eq. (2.28) to d dimensions reads

$$H = -J \sum_{\langle i,j \rangle} b_i^\dagger b_j + \frac{U}{2} \sum_i n_i(n_i - 1) - \mu \sum_i n_i \,, \qquad (2.33)$$

where the first sum goes only over nearest neighbours and we included the chemical potential μ which allows us to study the quantum statistics in the grand-canonical ensemble.

The reasoning illustrated above for 1 dimension can be applied also in d dimensions and a superfluid and a Mott-insulator phase can be identified. The two phases are distinguished by the superfluid order parameter $\Phi = \langle b_i \rangle$. In the Mott-insulator phase the eigenstates of the Hamiltonian are number states with well defined atom number at each site and hence $\Phi = 0$. In the superfluid phase each atom is delocalized over the whole lattice and the order parameter Φ takes on a nonzero value.

2.3.1 Mean-Field Treatment

When applying a mean-field treatment to a lattice Hamiltonian such as Eq. (2.28) one tries to find the best possible sum of single-site Hamiltonians which models its properties. The effects of neighbouring sites is taken into account by an effective field, which in the case of Eq. (2.28) is given by the superfluid order parameter Φ [68]. In the following we derive the mean-field approximation of Eq. (2.28) for d dimensions which allows us to obtain an analytical result for the quantum phase transition from Mott-insulator to superfluid state. Such treatment yields good quantitative results in

2 and 3 dimensions but not in 1 dimension, where the effects of quantum fluctuations cannot be treated by an effective field [68, 75].

In order to develop a mean field treatment we start from Hamiltonian Eq. (2.33) and introduce the operator

$$\beta_i = b_i - \Phi, \tag{2.34}$$

giving the fluctuations from the mean-field value. Substituting Eq. (2.34) in Eq. (2.33) leads to the term

$$H = \sum_i h_i - J \sum_{\langle i,j \rangle} \beta_i^\dagger \beta_j. \tag{2.35}$$

The on-site Hamiltonian h_i is given by

$$h_i = \frac{U}{2} n_i(n_i - 1) - \mu n_i + zJ\Phi^2 - zJ\Phi(b_i^\dagger + b_i), \tag{2.36}$$

where z is the number of nearest neighbours. In the mean-field approximation one neglects the nonlocal hopping term in Eq. (2.35) such that the full mean-field Hamiltonian becomes

$$H^{MF} = \sum_i h_i. \tag{2.37}$$

In this form the solution of the complete Hamiltonian Eq. (2.33) is reduced to the diagonalization of h_i for a single site. The eigenfunctions of H^{MF} are given by a tensor product of eigenfunctions for each site. The mean-field approximation thus neglects correlations between different sites. For small values of Φ we can split the mean-field Hamiltonian M^{MF} into an unperturbed part and a small perturbation

$$H_0^{MF} = \frac{U}{2} n_i(n_i - 1) - \mu n_i + O(\Phi^2), \tag{2.38}$$

$$V^{MF} = -zJ\Phi(b_i^\dagger + b_i). \tag{2.39}$$

In order to derive a mean-field equation describing the system close to the phase transition we calculate the expectation value $\langle b_l \rangle$ in perturbation theory. In first order perturbation theory the density operator ρ^{MF} reads [76]

$$\rho^{MF} = \frac{1}{Z} e^{-\beta H^{MF}} = \frac{1}{Z} e^{-\beta H_0^{MF}} - \frac{1}{Z} \int_0^\beta d\tau\, e^{-(\beta-\tau)H_0^{MF}} V^{MF} e^{-\tau H_0^{MF}} + O(\Phi^2), \tag{2.40}$$

where $Z = \text{tr}\left\{e^{-\beta H^{MF}}\right\} = \text{tr}\left\{e^{-\beta H_0^{MF}}\right\} + O(\Phi^2)$ is the partition sum and $\beta = \frac{1}{k_B T}$. Therefore the order parameter can be written as

$$\begin{aligned}
\Phi &= \langle b_i \rangle = \frac{1}{Z} \text{tr}\{e^{-\beta H^{MF}} b_i\} \\
&= \frac{Jz\Phi}{Z} \text{tr}\left\{\int_0^\beta d\tau\, e^{-(\beta-\tau)H_0^{MF}} b_i^\dagger e^{-\tau H_0^{MF}} b_i\right\} + O(\Phi^2).
\end{aligned} \tag{2.41}$$

2.3 Bose-Hubbard Hamiltonian

In the zero temperature limit $\beta \to \infty$ Eq. (2.41) becomes

$$\Phi = zJ\Phi \sum_{|\psi\rangle}(n^{|\psi\rangle}+1)\left[\frac{\exp[-\beta(H_0^{MF|\psi\rangle} - E_0)]}{H_0^{MF|\psi+1,i\rangle} - H_0^{MF|\psi\rangle}} - \frac{\exp[-\beta(H_0^{MF|\psi+1,i\rangle} - E_0)]}{H_0^{MF|\psi+1,i\rangle} - H_0^{MF|\psi\rangle}}\right], \tag{2.42}$$

where E_0 is the ground state energy. We introduced the writting $A^{|\psi\rangle} = \langle\psi|A|\psi\rangle$ and $|\psi+1,i\rangle$ is the state which results from adding to the state $|\psi\rangle$ one atom at site i. Since E_0 is the ground state energy one finds in the limit $\beta \to \infty$ only two contributions to the sum, namely if either $|\psi\rangle$ or $|\psi+1,i\rangle$ is the ground state. In the Mott-insulator phase there is a well defined number of atoms g at each site, such that

$$H_0^{MF|\psi+1,i\rangle} - H_0^{MF|\psi\rangle} = Ug - \mu. \tag{2.43}$$

Using Eq. (2.43) in Eq. (2.42) one finds the following condition for the order parameter

$$\Phi\left(1 - zJ\left[\frac{g+1}{Ug-\mu} + \frac{g}{\mu - U(g-1)}\right]\right) = 0. \tag{2.44}$$

The solution of Eq. (2.44) is $\Phi = 0$ unless the expression inside the parentheses vanishes, corresponding to the condition

$$zJ = \frac{(Ug-\mu)(\mu - U(g-1))}{\mu + U}. \tag{2.45}$$

Equation (2.45) allows one to calculate the chemical potential μ as a function of zJ for constant filling g, hence determining the boundary between the Mott-insulator phase with filling g and the superfluid phase. We note that Eq. (2.45) has only solutions for $zJ > 0$ if

$$g - 1 \leq \frac{\mu}{U} \leq g, \tag{2.46}$$

which leads to separate parameter regions for given filling g, where the system is in the Mott-insulator phase. Figure (2.6a) displays the phase diagram of the Bose-Hubbard Hamiltonian as a function of chemical potential μ and hopping zJ in units of the on-site interaction strength U, evaluated by means of the mean-field approximation. A lobe structure appears where within each lobe the system is in the Mott-insulator state and the number of particles in each well is a fixed integer number. Outside of the lobes the system is in the superfluid regime. The value of $\tilde{J}_c = zJ_c/U_c$ for the ratio between tunneling matrix element and on-site interaction at the tips of the lobes can be determined by maximizing Eq. (2.45) for zJ/U, which leads to the relation

$$\tilde{J}_c = 1 + 2g - 2(g+g^2)^{\frac{1}{2}}. \tag{2.47}$$

Equation (2.47) determines the critical ratio $\tilde{J}_c$ between hopping and on-site interaction for the quantum phase transition at constant density. In the superfluid phase the order parameter has to be evaluated numerically by diagonalizing H^{MF}. In order to diagonalize the mean-field Hamiltonian we expand it in the number state basis up to a maximal occupation number n_M and then minimize the ground state energy as a function of Φ. We then double n_M and repeat this process till the relative ground state energy has converged to the ratio $\frac{E_0(n_M)}{E_0(2n_M)} \leq 10^{-5}$. Comparison of the results obtained

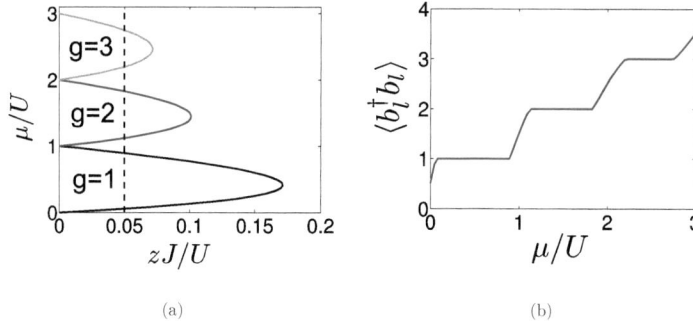

(a) (b)

Figure 2.6: a) Phase diagram of the Bose-Hubbard Hamiltonian in the mean-field approximation. The y-axis shows the chemical potential μ in units of the on-site interaction strength U and the x-axis the hopping J in units of U/z, where z is the number of nearest neighbours of a given site. The boundaries of the Mott lobes are found from Eq. (2.45). b) Average number occupation $\langle b_l^\dagger b_l \rangle$ as a function of chemical potential μ in units of the on-site energy U at fixed hopping $zJ = 0.05U$, corresponding to the dashed line in the phase diagram in (a).

from the approximate Hamiltonian Eq. (2.37) with quantum Monte Carlo simulations shows that the mean-field approximation gives good quantitative results for 2 and 3-dimensional systems while in 1-dimensional systems large deviations are observed [75].

We finally remark that the model here presented is valid for homogeneous systems while in typical experiments the atoms are confined by an additional harmonic potential. This fact can be theoretically taken into account by applying the local density approximation, which is valid if the external potential only varies slightly over the typical length scales of the system. In this case one can take the solutions for the homogeneous system with a spatially varying chemical potential which mimics the external potential. This means by going from the border of the system towards the center of the trap one moves in the phase diagram from low to higher values of the chemical potential. In Fig. (2.6b)

2.3 Bose-Hubbard Hamiltonian

the average number of particles in each well $\langle b_l^\dagger b_l \rangle$ is plotted as a function of chemical potential for fixed zJ/U. For values of μ which correspond to the Mott-insulator regime the average number of atoms at each site takes on well defined integer numbers $\langle b_l^\dagger b_l \rangle = j$ with $j = 1, 2, 3, \ldots$, thus showing that each lattice site is occupied by a fixed number of atoms. If the system is in the superfluid state, $\langle b_l^\dagger b_l \rangle$ takes on noninteger values and increases monotonically with the chemical potential μ, thus giving nonzero number fluctuations at each site.

2.3.2 Bogoliubov Expansion

We now study the dynamics and the spectrum of Hamiltonian Eq. (2.33) in the superfluid regime. Since most of the atoms will be in the lowest energy eigenstate we can use a Bogoliubov description for the system by splitting the atomic operators according to Eq. (2.34) and treating the effect of the operator β_j as a small perturbation to the mean-field equations for the order parameter Φ. Splitting Hamiltonian Eq. (2.33) into the different orders in β_j we find[2]

$$H = H_0 + H_1 + H_2 + H_3 + H_4, \qquad (2.48)$$

with

$$H_0 = \sum_l \left[-zJ - \mu + \frac{U}{2}|\Phi|^2 \right] |\Phi|^2, \qquad (2.49)$$

$$H_1 = \sum_l \beta_l^\dagger \left[U|\Phi|^2 - zJ - \mu \right] \Phi + \text{H.c.}, \qquad (2.50)$$

$$H_2 = -J \sum_{\langle l,m \rangle} \beta_l^\dagger \beta_m + \frac{U}{2} \sum_l \left[4|\Phi|^2 \beta_l^\dagger \beta_l + \Phi^2 \beta_l^{\dagger 2} + \bar{\Phi}^2 \beta_l^2 \right] - \mu \sum_l \beta_l^\dagger \beta_l, \quad (2.51)$$

$$H_3 = U \sum_l \bar{\Phi} \beta_l^\dagger \beta_l^2 + \text{H.c.}, \qquad (2.52)$$

$$H_4 = \frac{U}{2} \sum_l \beta_l^{2\dagger} \beta_l^2. \qquad (2.53)$$

Minimizing the scalar equation for the energy found from Hamiltonian H_0 with respect to Φ one finds

$$\mu = -zJ + U|\Phi|^2, \qquad (2.54)$$

which is the discrete form of the Gross-Pitaevskii equation [61] for a homogeneous system. If the order parameter satisfies Eq. (2.54) Hamiltonian H_1 vanishes identically. Hamiltonian H_0 only describes the contribution of the condensate, whereas H_2 takes

---
[2]In the following we take for convenience the on-site interaction part in the form $\frac{U}{2} \langle b_l^\dagger b_l^\dagger b_l b_l \rangle$.

into account the leading order effects of the non condensate fraction. Hamiltonian H_2 can be diagonalized by the Bogoliubov transformation

$$\beta_l = \sum_s u_l^s \alpha_s - \bar{v}_l^s \alpha_s^\dagger, \qquad (2.55a)$$

$$\alpha_s = \sum_l \bar{u}_l^s \beta_l + \bar{v}_l^s \beta_l^\dagger, \qquad (2.55b)$$

where u_l^s, v_l^s are complex numbers and α_s, $\alpha_s^\dagger$ are annihilation and creation operators of a quasi-particle in the Bogoliubov mode s. Equation (2.55b) is the inverse transformation of Eq. (2.55a). The amplitudes u_l^s and v_l^s obey the following constraints

$$\sum_n u_n^s v_n^r - u_n^r v_n^s = 0,$$

$$\sum_n \bar{u}_n^r u_n^s - \bar{v}_n^r v_n^s = \delta_{r,s},$$

$$\sum_s u_n^s \bar{v}_k^s - \bar{v}_n^s u_k^s = 0,$$

$$\sum_s u_n^s \bar{u}_k^s - \bar{v}_n^s v_k^s = \delta_{k,n}. \qquad (2.56)$$

It is easily verified that the Bogoliubov transformation is canonical, so that the quasi-particle operators α_s obey the usual bosonic commutation relations $\left[\alpha_s, \alpha_{s'}^\dagger\right] = \delta_{s,s'}$. Hamiltonian H_2 can be written in the following form

$$H_2 = \sum_{l,m} \mathcal{L}_{l,m} \beta_l^\dagger \beta_m + \mathcal{M}_{l,m} \beta_l^\dagger \beta_m^\dagger + \text{H.c.}, \qquad (2.57)$$

where the coefficients read

$$\mathcal{L}_{l,m} = -J \sum_{\langle n,k \rangle} \delta_{n,l} \delta_{m,k}/2 + \delta_{l,m}(2U|\Phi|^2 - \mu)/2,$$

$$\mathcal{M}_{l,m} = U\Phi^2 \delta_{l,m}/2.$$

Inserting Eq. (2.55a) into H_2 yields a "nondiagonal" part containing products of $\alpha_s \alpha_r$ and $\alpha_s^\dagger \alpha_r^\dagger$ which vanishes due to the inverse transformation condition Eq. (2.56). The remaining part reads

$$H_2^{diag} = \sum_{s,r} \sum_{l,m} \mathcal{L}_{l,m} \left(\bar{u}_l^s u_m^r \alpha_s^\dagger \alpha_r + v_l^s \bar{v}_m^r \alpha_s \alpha_r^\dagger \right) - \mathcal{M}_{l,m} \left(\bar{u}_l^s v_m^r \alpha_s^\dagger \alpha_r^\dagger + v_l^s \bar{u}_m^r \alpha_s \alpha_r^\dagger \right) + \text{H.c.}. \qquad (2.58)$$

Equations (2.58) takes the form $H_2^{diag} = \sum_s \hbar\Omega_s \alpha_s^\dagger \alpha_s + C_s$, with C_s being some mode dependent c-number, if the amplitudes u_l^s and v_l^s satisfy

$$\sum_{l,m} \mathcal{L}_{l,m} \left(\bar{u}_l^s u_m^r + v_l^s \bar{v}_m^r \right) - \mathcal{M}_{l,m} \left(\bar{u}_l^s v_m^r + v_l^s \bar{u}_m^r \right) = \frac{\hbar\Omega_s}{2} \sum_l \bar{u}_l^s u_l^r - \bar{v}_l^s v_l^r, \qquad (2.59)$$

2.3 Bose-Hubbard Hamiltonian

where we used Eq. (2.56). This leads to the so called Bogoliubov-de Gennes (BdG) equations [61]

$$2\sum_m \mathcal{L}_{l,m} u_m^s - \mathcal{M}_{l,m} v_m^s = \hbar\Omega_s u_l^s, \tag{2.60a}$$

$$2\sum_m \mathcal{L}_{l,m} v_m^s - \bar{\mathcal{M}}_{l,m} u_m^s = -\hbar\Omega_s v_l^s. \tag{2.60b}$$

If Eqs. (2.60) are satisfied Hamiltonian H_2 takes the form

$$H_2 = \sum_s \left[\hbar\Omega_s \left(\alpha_s^\dagger \alpha_s + \frac{1}{2}\right) - \sum_l \mathcal{L}_{ll}\right]. \tag{2.61}$$

The solution of Eq. (2.60) is usually involved but simple solutions can be found for a homogeneous system. There the Bogoliubov modes are characterized by their quasi-momentum $\mathbf{p}$ and read

$$u_l^\mathbf{p} = \frac{1}{\sqrt{M}} e^{i\mathbf{p}\cdot\mathbf{x}_l} u_\mathbf{p}, \tag{2.62a}$$

$$v_l^\mathbf{p} = \frac{1}{\sqrt{M}} e^{i\mathbf{p}\cdot\mathbf{x}_l} v_\mathbf{p}, \tag{2.62b}$$

where Eqs. (2.56) lead to the conditions $|u_\mathbf{p}|^2 - |v_\mathbf{p}|^2 = 1$ and $u_\mathbf{p} v_{-\mathbf{p}} = u_{-\mathbf{p}} v_\mathbf{p}$. The components of the quasi-momentum $\mathbf{p} = \sum_{j=x,y,z} p_j \hat{\mathbf{e}}_j$ take the discrete values

$$p_j = \frac{2\pi}{d_j M_j} n_j \quad \text{with} \quad n_j = -M_j, 1-M_j \ldots M_j - 1 \tag{2.63}$$

and the BdG-equations reduce to the following 2×2 eigenvalue problem

$$\begin{pmatrix} \tilde{\epsilon}_\mathbf{p} + U|\Phi|^2 & -U\Phi^2 \\ U\bar\Phi^2 & -\tilde{\epsilon}_\mathbf{p} - U|\Phi|^2 \end{pmatrix} \begin{pmatrix} u_\mathbf{p} \\ v_\mathbf{p} \end{pmatrix} = \hbar\Omega_\mathbf{p} \begin{pmatrix} u_\mathbf{p} \\ v_\mathbf{p} \end{pmatrix}, \tag{2.64}$$

where Eq. (2.54) has been used and

$$\tilde{\epsilon}_\mathbf{p} = 4J \sum_{j=x,y,z} \sin^2\left(\frac{d_j p_j}{2}\right). \tag{2.65}$$

The solution for the spectrum and the Bogoliubov amplitudes is given by

$$\hbar^2 \Omega_\mathbf{p}^2 = \tilde{\epsilon}_\mathbf{p}^2 + 2U n_0 \tilde{\epsilon}_\mathbf{p}, \tag{2.66}$$

and

$$|u_\mathbf{p}|^2 = \frac{\tilde{\epsilon}_\mathbf{p} + U n_0 + \hbar\Omega_\mathbf{p}}{2\hbar\Omega_\mathbf{p}}, \tag{2.67a}$$

$$|v_\mathbf{p}|^2 = \frac{\tilde{\epsilon}_\mathbf{p} + U n_0 - \hbar\Omega_\mathbf{p}}{2\hbar\Omega_\mathbf{p}}, \tag{2.67b}$$

$$u_\mathbf{p} \bar{v}_\mathbf{p} = \frac{U\Phi^2}{2\hbar\Omega_\mathbf{p}}, \tag{2.67c}$$

where $n_0 = |\Phi|^2$ is the condensate (superfluid) fraction. We note that $\tilde{\epsilon}_{\mathbf{p}}$ is the energy of a non interacting particle in the lattice. By replacing it with the free-space energy $\tilde{\epsilon}_{\mathbf{p}} \to \mathbf{p}^2/2m$ we recover in Eq. (2.66) the dispersion relation for a weakly-interacting dilute Bose gas in free space [61].

Equation. (2.61) is a general result which holds for every lattice-Hamiltonian that can be written in the form of Eq. (2.57). Hence an external potential or site dependent hopping strength is easily implemented in this formalism. However for inhomogeneous systems solving the BdG-equations Eqs. (2.60) becomes much more demanding than in the case treated here.

2.3.3 Particle-Hole Expansion

We now consider the Mott-insulator regime and treat the hopping J as a perturbation. In the limit of large filling the excitations arising from adding or subtracting a particle to the system are degenerate. This introduces an additional symmetry which allows one to obtain analytical formulas for the eigenstates and eigenenergies of the Bose-Hubbard Hamiltonian up to first order in J for large filling [77, 21].

For vanishing hopping, the ground state of Hamiltonian (2.28) is the Mott-insulator state with all lattice sites equally occupied with (integer) filling factor $g = N/M$ as given in Eq. (2.32). The corresponding ground state energy for $J = 0$ is easily found and reads $E_0^0 = MUg(g-1)/2 - Mg\mu$. The lowest-lying excitations take the form

$$|\psi_{n,m}^{(0)}\rangle = \frac{b_n^\dagger b_m}{\sqrt{g(g+1)}}|\psi_0^{(0)}\rangle, \quad (2.68)$$

where one particle and one hole are created at site n and m, respectively, with energy $E_1^0 = E_0^0 + U$. These states form a degenerate subspace of dimension $M(M-1)$. This degeneracy is lifted for finite values of the hopping J. The corrections due to a non-vanishing but small value of tunneling are evaluated using perturbation theory. Including the first-order correction, the ground state now reads

$$|\psi_0^{(1)}\rangle = \left(1 - \frac{J^2}{U^2}Mg(g+1)\right)|\psi_0^{(0)}\rangle + \frac{J}{U}\sqrt{2Mg(g+1)}|S\rangle, \quad (2.69)$$

where $|S\rangle = \frac{1}{\sqrt{2M}}\sum_n \left(|\psi_{n,n+1}^{(0)}\rangle + |\psi_{n,n-1}^{(0)}\rangle\right)$ is the normalized state of adjacent particle-hole excitations, while the term in second order in J warrants normalization of state (2.69). The corresponding energy is $E_0 = E_0^0 + O(J^2)$. The lowest-lying excitations are determined using degenerate perturbation theory within the subspace of single particle-hole excitations,

$$|\psi_{[i]}^{(0)}\rangle = \sum_{n,m} c_{n,m}^{[i]}|\psi_{n,m}^{(0)}\rangle, \quad (2.70)$$

2.3 Bose-Hubbard Hamiltonian

where the coefficients $c_{n,m}^{[i]}$ fulfill the normalization condition $\sum_{n,m}|c_{n,m}^{[i]}|^2$ and satisfy the equations

$$(g+1)(c_{n+1,m}^{[i]} + c_{n-1,m}^{[i]}) + g(c_{n,m+1}^{[i]} + c_{n,m-1}^{[i]}) = A_i c_{n,m}^{[i]}, \quad (2.71)$$

with periodic boundary conditions

$$c_{n+M,m}^{[i]} = c_{n,m+M}^{[i]} = c_{n,m}^{[i]}, \quad (2.72)$$

$$c_{n,n}^{[i]} = 0. \quad (2.73)$$

The term A_i in Eq. (2.71) is the first-order correction to the corresponding energy, $E_i = E_0 + U - JA_i + O(J^2)$. Using the *Ansatz* [77]

$$c_{n,m}^{[r,s,t]} = \frac{1}{\mathcal{N}} \sin \alpha r(n-m) e^{\imath \alpha s n} e^{\imath \alpha t m} \quad (2.74)$$

with

$$\alpha = \frac{\pi}{M}, \quad (2.75)$$

the following equation for the parameters $[r,s,t]$ is found

$$2 \sin \alpha r(n-m) \cos \alpha r \left[(g+1) \cos \alpha s + g \cos \alpha t\right]$$
$$+ 2\imath \cos \alpha r(n-m) \sin \alpha r \left[(g+1) \sin \alpha s - g \sin \alpha t\right] = A \sin \alpha r(n-m). \quad (2.76)$$

This equation is satisfied if the condition

$$s = \arcsin\left[\frac{g}{g+1} \sin t\right] \quad (2.77)$$

is fulfilled. In general the coefficients $c_{n,m}^{[r,s,t]}$ obtained from Eq. (2.77) violate the periodic boundary condition in Eq. (2.72). However for large $g \gg 1$ we can take $s \approx t$ and an analytic solution for the coefficients $c_{n,m}^{[r,s]}$ can be obtained. Taking $s = t$ introduces a symmetry between particle and hole excitations which is strictly correct only when $g \to \infty$. The coefficients, evaluated in this limit, read

$$c_{n,m}^{[r,s]} = \frac{\sqrt{2}}{M} \begin{cases} \sin[\alpha r|n-m|] \, e^{\imath \alpha s(n+m)} & \text{for } r+s \text{ odd,} \\ \sin[\alpha r(n-m)] \, e^{\imath \alpha s(n+m)} & \text{for } r+s \text{ even,} \end{cases} \quad (2.78)$$

with $\alpha = \frac{\pi}{M}$, $s = 0, 1 \ldots M-1$ and $r = 1, 2 \ldots M-1$. Correspondingly, the lowest-lying excitations and their energy are at first order in J and for $g \gg 1$ given by

$$|\psi_{[r,s]}^{(1)}\rangle = \begin{cases} \frac{1}{\mathcal{N}_r}\left(\sum_{n,m} \left(c_{n,m}^{[r,0]}|\psi_{n,m}\rangle\right) - \frac{J}{U}\sqrt{8g(g+1)} \sin \alpha r |\psi_0^{(0)}\rangle\right) & \text{if } s=0, r \text{ odd,} \\ \sum_{n,m} c_{n,m}^{[r,s]}|\psi_{n,m}\rangle & \text{otherwise,} \end{cases} \quad (2.79a)$$

$$E_{r,s} = E_0^0 + U - 2J(2g+1) \cos \alpha r \cos \alpha s + O(J^2), \quad (2.79b)$$

where $\mathcal{N}_r$ is a normalization factor. Note that the states $|\psi_{[r,s]}^{(1)}\rangle$ contain a correction proportional to the ground state $|\psi_0^{(0)}\rangle$ which warrants the orthonormality of the new basis $\{|\psi_0^{(1)}\rangle, |\psi_{[r,s]}^{(1)}\rangle\}$. Figure (2.7) displays the analytic solution of the energy corrections to the first excited states $\Delta E_n^1 = -2J(2g+1)\cos\alpha r \cos\alpha s$ where $n = (r,s)$ versus the numerical solutions obtained by diagonalizing the hopping term V in the restricted Hilbert space of at most one particle-hole excitation. We see that the analytical solutions Eqs. (2.79) agree well with the numerical solution, where for a filling factor of $g = 4$ almost no deviations can be observed.

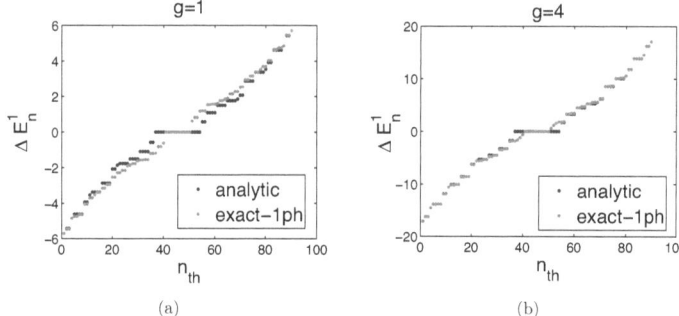

Figure 2.7: Comparison of the first order corrections of the 1-ph excitations between the analytic solution and numerical diagonalization of V within the 1-ph- subspace. The number of sites is $M = 10$ and the energies are in units of J.

2.3.4 Random phase approximation

Within various methods of solution of many-body problems Green functions have the advantage that the calculation of expectation values of time dependent operators is reduced to solve coupled differential equations of complex functions. To get an exact solution, however one has in general to solve an infinite set of coupled differential equations. The so called method based on the random phase approximation (RPA) decouples some of the equations in order to get a closed set of equations. The RPA is widely used in condensed matter theory and solid state physics. In this section we discuss its application to the Bose-Hubbard model, which yields approximate expressions for the spectrum in both the superfluid and Mott-insulator regime.

In order to proceed it is convenient to expand Hamiltonian Eq. (2.35) in eigenstates

2.3 Bose-Hubbard Hamiltonian

of the on-site Hamiltonian Eq. (2.37) $H^{MF}|l,\alpha\rangle = E_\alpha|l,\alpha\rangle$, thereby obtaining

$$H = \sum_l E_\alpha L^l_{\alpha,\alpha} - J \sum_{<l,m>} \sum_{\alpha,\alpha'\beta\beta'} T^{l,m}_{\alpha,\alpha'\beta\beta'} L^l_{\alpha,\alpha'} L^m_{\beta,\beta'}, \qquad (2.80)$$

with $L^l_{\alpha,\beta} = |l,\alpha\rangle\langle l,\beta|$ being the standard basis operators [78] and

$$T^{l,m}_{\alpha,\alpha'\beta\beta'} = \langle l,\alpha|\beta_l^\dagger|l,\alpha'\rangle\langle m,\beta|\beta_m|m,\beta'\rangle \qquad (2.81)$$

is the coupling term. For a homogeneous system the eigenenergies $E^l_\alpha = E_\alpha$ and coupling terms $T^{l,m}_{\alpha,\alpha'\beta\beta'} = T_{\alpha,\alpha'\beta\beta'}$ do not depend on position. We introduce the retarded Green function for two operators A_l and B_m which act only at site l and m respectively

$$G^{l,m}_{AB}(t-t') = -\frac{i}{\hbar}\Theta(t-t')\langle[A(t),B(t')]\rangle := \langle\langle A(t);B(t')\rangle\rangle. \qquad (2.82)$$

Expansion of the Green function in standard basis operators leads to [78]

$$G^{l,m}_{AB}(t-t') = \sum_{\alpha,\alpha'\beta\beta'} \langle l,\alpha|A_l|l,\alpha'\rangle\langle m,\beta|B_m|m,\beta'\rangle G^{l,m}_{\alpha\alpha'\beta\beta'}(t-t'), \qquad (2.83)$$

with

$$G^{lm}_{\alpha\alpha'\beta\beta'}(t) = \langle\langle L^l_{\alpha\alpha'}(t); L^m_{\beta\beta'}(0)\rangle\rangle. \qquad (2.84)$$

The equation of motion for the Green function in frequency space reads

$$(\hbar\omega + E_\alpha - E_{\alpha'})G^{lm}_{\alpha\alpha'\beta\beta'}(\omega) = \delta_{l,m}\delta_{\alpha'\beta}\delta_{\alpha\beta'}D_{\alpha\alpha'} - J\sum_{j=l\pm}\sum_{\gamma\gamma'\nu} \qquad (2.85)$$

$$\left(\tilde{T}_{\alpha'\nu\gamma\gamma'}\left\langle\left\langle L^j_{\gamma\gamma'}L^l_{\alpha\nu}; L^m_{\beta\beta'}\right\rangle\right\rangle \right.$$

$$\left. -\tilde{T}_{\nu\alpha\beta\beta'}\left\langle\left\langle L^j_{\gamma\gamma'}L^l_{\nu\alpha'}; L^m_{\beta\beta'}\right\rangle\right\rangle\right),$$

with $D_{\alpha\beta} = \langle L^l_{\alpha\alpha}\rangle - \langle L^l_{\beta\beta}\rangle$ which is independent of l and $\tilde{T}_{\alpha'\nu\gamma\gamma'} = T_{\alpha'\nu\gamma\gamma'} + T_{\gamma\gamma'\alpha'\nu}$. The notation $j = l\pm$ means that the sum over j goes only over nearest neighbours of l. Within the RPA one decouples the three point correlators [78]

$$\left\langle\left\langle L^j_{\gamma\gamma'}L^l_{\nu\alpha'}; L^m_{\beta\beta'}\right\rangle\right\rangle = \langle L^j_{\gamma\gamma'}\rangle\left\langle\left\langle L^l_{\nu\alpha'}; L^m_{\beta\beta'}\right\rangle\right\rangle + \langle L^l_{\nu\alpha'}\rangle\left\langle\left\langle L^j_{\gamma\gamma'}; L^m_{\beta\beta'}\right\rangle\right\rangle, \qquad (2.86)$$

which leads to the approximate equation of motion

$$(\hbar\omega + E_{\alpha\alpha'})G_{\alpha\alpha'\beta\beta'}(\mathbf{k},\omega) = \delta_{\alpha'\beta}\delta_{\alpha\beta'}D_{\alpha\alpha'} + D_{\alpha\alpha'}\epsilon_\mathbf{k}\sum_{\nu\mu}\tilde{T}_{\alpha'\alpha\nu\mu}G_{\nu\mu\beta\beta'}(\mathbf{k},\omega). \qquad (2.87)$$

We introduced $E_{\alpha\alpha'} = (E_\alpha - E_{\alpha'})$ and

$$\epsilon_{\mathbf{k}} = -2J \sum_{j=(x,y,z)} \cos(k_j d_j), \qquad (2.88)$$

$$G_{\alpha\alpha'\beta\beta'}(\mathbf{k},\omega) = \sum_l e^{-i\mathbf{k}\cdot(\mathbf{x}_l - \mathbf{x}_m)} G_{\alpha\alpha'\beta\beta'}^{lm}(\omega). \qquad (2.89)$$

Comparison with Eq. (2.65) shows that the $\epsilon_{\mathbf{k}} = \tilde{\epsilon}_{\mathbf{k}} - zJ$. The solution of Eq. (2.87) allows the calculation of any correlation function of two operators at two spatial points. We now proceed and solve Eq. (2.87) considering the Green functions

$$G_{bb^\dagger}(\mathbf{k},\omega) = \int_{-\infty}^{\infty} dt\, e^{i\omega_+ t} \sum_l e^{-i\mathbf{k}\cdot(\mathbf{x}_l - \mathbf{x}_m)} \left\langle\!\left\langle b_l(t) b_m^\dagger(0) \right\rangle\!\right\rangle, \qquad (2.90a)$$

$$G_{b^\dagger b^\dagger}(\mathbf{k},\omega) = \int_{-\infty}^{\infty} dt\, e^{i\omega_+ t} \sum_l e^{-i\mathbf{k}\cdot(\mathbf{x}_l - \mathbf{x}_m)} \left\langle\!\left\langle b_l^\dagger(t) b_m^\dagger(0) \right\rangle\!\right\rangle, \qquad (2.90b)$$

which can be expanded according to Eq. (2.83). The notation $\omega_+ = \lim_{\eta \to 0+} \omega + i\eta$ in the definition of the Fourier transform is meant to remind that the perturbation is taken to be zero at $t = -\infty$ and switched on adiabatically. Substituting Eq. (2.90) into the equations of motion Eq. (2.87) we find

$$G_{bb^\dagger}(\mathbf{k},\omega) = A_1(\omega) + \epsilon_{\mathbf{k}} A_1(\omega) G_{bb^\dagger}(\mathbf{k},\omega) + \epsilon_{\mathbf{k}} A_2(\omega) G_{b^\dagger b^\dagger}(\mathbf{k},\omega), \qquad (2.91a)$$

$$G_{b^\dagger b^\dagger}(\mathbf{k},\omega) = A_3(\omega) + \epsilon_{\mathbf{k}} A_3(\omega) G_{bb^\dagger}(k,\omega) + \epsilon_{\mathbf{k}} A_1(-\omega) G_{b^\dagger b^\dagger}(\mathbf{k},\omega), \qquad (2.91b)$$

with

$$A_1(\omega) = \sum_\alpha \frac{\langle l,\alpha | b_l^\dagger | l, 0\rangle \langle l, 0 | b_l | l, \alpha\rangle}{\hbar\omega + E_{0\alpha}} - \frac{\langle l, 0 | b_l^\dagger | l, \alpha\rangle \langle l, \alpha | b_l | l, 0\rangle}{\hbar\omega + E_{\alpha 0}}, \qquad (2.92a)$$

$$A_2(\omega) = \sum_\alpha \langle l, \alpha | b_l | l, 0\rangle \langle l, 0 | b_l | l, \alpha\rangle \frac{2 E_{\alpha 0}}{\hbar^2 \omega^2 - E_{\alpha 0}^2}, \qquad (2.92b)$$

$$A_3(\omega) = \sum_\alpha \langle l, \alpha | b_l^\dagger | l, 0\rangle \langle l, 0 | b_l^\dagger | l, \alpha\rangle \frac{2 E_{\alpha 0}}{\hbar^2 \omega^2 - E_{\alpha 0}^2}. \qquad (2.92c)$$

From Eq. (2.91) one finds the compact expressions [79]

$$G_{bb^\dagger}(\mathbf{k},\omega) = \frac{\Pi(\mathbf{k},\omega)}{1 - \epsilon_{\mathbf{k}} \Pi(\mathbf{k},\omega)}, \qquad (2.93)$$

$$G_{b^\dagger b^\dagger}(\mathbf{k},\omega) = \frac{A_3(\omega)}{(1 - \epsilon_{\mathbf{k}} A_1(-\omega))(1 - \epsilon_{\mathbf{k}} \Pi)}, \qquad (2.94)$$

where

$$\Pi(\mathbf{k},\omega) = A_1(\omega) + \frac{\epsilon_{\mathbf{k}} |A_2(\omega)|^2}{1 - \epsilon_{\mathbf{k}} A_1(-\omega)}. \qquad (2.95)$$

2.3 Bose-Hubbard Hamiltonian

Now we introduce the spectral functions

$$A_{PQ}(\mathbf{k},\omega) = \int dt e^{i\omega t} \sum_{l} e^{-i\mathbf{k}\cdot(\mathbf{r}_l-\mathbf{r}_m)} \langle[P_l(t),Q_m(0)]\rangle \quad (2.96)$$

for two operators P and Q, which are related to the Green functions via the relation

$$A_{PQ}(\mathbf{k},\omega) = -2\hbar\mathrm{Im}\big(G_{PQ}(\mathbf{k},\omega+i\eta)\big). \quad (2.97)$$

The relation Eq. (2.97) can be easily verified from the Lehman representation of the two functions [80]. The spectral density $A_{bb^\dagger}(\mathbf{k},\omega)$ allows one to calculate various physical quantities, such as the momentum distribution, which is given by

$$n_\mathbf{k} = |\langle b\rangle|^2 M\delta_{\mathbf{k},0} + \frac{1}{2\pi}\int_{-\infty}^{\infty} d\omega \frac{A_{bb^\dagger}(\mathbf{k},\omega)}{e^{\hbar\omega\beta}-1}. \quad (2.98)$$

Here $n_\mathbf{k}$ is the expectation value for the number of atoms at quasi-momentum $\mathbf{k}$ and we omitted to write the index l in the first term since for a homogeneous lattice $\langle b_l\rangle$ is independent on the lattice site. The first term in Eq. (2.98) describes the condensate of atoms in the lowest momentum state of the lattice and is hence nonvanishing only in the superfluid regime. We also note that the spectral density $A_{bb^\dagger}(\mathbf{k},\omega)$ obeys the sum rule

$$\int_{-\infty}^{\infty} d\omega A_{bb^\dagger}(\mathbf{k},\omega) = 2\pi. \quad (2.99)$$

In the Mott insulator phase the states $|l,\alpha\rangle$ are all Fock states and the Green function $G_{bb^\dagger}(\mathbf{k},\omega)$ can be calculated analytically [79]

$$G_{bb^\dagger}(\mathbf{k},\omega) = \frac{(g+1)(\hbar\omega+\Delta_-) - g(\hbar\omega-\Delta_+)}{(\hbar\omega-\Delta_+)(\hbar\omega+\Delta_-) - \epsilon_\mathbf{k}(\hbar\omega+U+\mu)}, \quad (2.100)$$

where $\Delta_\pm = E_{g\pm1} - E_g$ are the energy differences between a state with $g\pm1$ atoms and g atoms of the on-site Hamiltonian Eq. (2.37) and are given by

$$\Delta_+ = Ug - \mu, \quad (2.101)$$
$$\Delta_- = \mu - U(g-1). \quad (2.102)$$

The energies $\Delta_\pm$ are both positive as one can check by using Eq. (2.46). The superfluid Mott-insulator phase transition is determined by requiring Green function Eq. (2.100) to have a pole at $\omega=0$ for $\mathbf{k}\to 0$, which is equivalent to condition Eq. (2.45). We note that Eq. (2.100) is non perturbative in the hopping J since an expansion of the denominator contains all orders of J. For zero tunneling $J=0$ we find

$$\hbar G_{bb^\dagger}(\mathbf{k},\omega) = \frac{g+1}{\omega-\Delta_+} - \frac{g}{\omega+\Delta_-}, \quad (2.103)$$
$$A_{bb^\dagger}(\mathbf{k},\omega) = 2\pi\left[(g+1)\delta(\omega-\Delta_+) - g\delta(\omega+\Delta_-)\right], \quad (2.104)$$

which yields a flat momentum distribution at zero temperature, $n_{\mathbf{k}} = g$. For finite tunneling J the Green function and spectral density can be written in the form

$$\hbar G_{bb^\dagger}(\mathbf{k},\omega) = \frac{(g+1)(\omega + \Delta_-) - g(\omega - \Delta_+)}{\omega_1 - \omega_2} \left(\frac{1}{\omega - \omega_1} - \frac{1}{\omega - \omega_2} \right), \quad (2.105a)$$

$$A_{bb^\dagger}(\mathbf{k},\omega) = 2\pi \frac{(g+1)(\omega + \Delta_-) - g(\omega - \Delta_+)}{\omega_1 - \omega_2} \left(\delta(\omega - \omega_1) - \delta(\omega - \omega_2) \right), \quad (2.105b)$$

with

$$\omega_{1,2} = \frac{1}{2} \left(U(2g-1) + \epsilon_{\mathbf{k}} - 2\mu \pm \sqrt{(\epsilon_{\mathbf{k}} + U)^2 + 4\epsilon_{\mathbf{k}} U g} \right), \quad (2.106)$$

where ω_1 (ω_2) corresponds to the $+$ ($-$) sign. The term in front of the square root becomes minimal when maximizing the chemical potential $\mu_{max} = Ug$, giving $\omega_1 > 0$ and $\omega_2 < 0$.

The momentum distribution at $T = 0$ in the Mott-insulator state is given by

$$n_{\mathbf{k}} = -\frac{1}{2\pi} \int_{-\infty}^{0} d\omega A_{bb^\dagger}(\mathbf{k},\omega) = \frac{(g+1)(\omega_2 + \Delta_-) - g(\omega_2 - \Delta_+)}{\omega_1 - \omega_2}, \quad (2.107)$$

which can be simplified to the form

$$n_{\mathbf{k}} = \frac{\epsilon_{\mathbf{k}} + U(2g+1)}{2\sqrt{(\epsilon_{\mathbf{k}} + U)^2 + 4\epsilon_{\mathbf{k}} U g}} - \frac{1}{2}. \quad (2.108)$$

Expanding Eq. (2.108) in powers of J/U one obtains

$$n_{\mathbf{k}} = g - 2\frac{\epsilon_{\mathbf{k}}}{U} g(g+1) + \frac{\epsilon_{\mathbf{k}}^2}{2U^2}(2g+1)(6g^2 + 6g + 1) + O\left(\frac{J^3}{U^3}\right), \quad (2.109)$$

which agrees up to first order to the result that can be found by using the perturbative result of Sec. (2.3.3). We note that while the sum rule Eq. (2.99) is perfectly satisfied, the normalization of the momentum distribution is instead only fulfilled up to first order in J/U,

$$\frac{1}{N} \sum_{\mathbf{k}} n_{\mathbf{k}} = 1 + O(J^2/U^2), \quad (2.110)$$

where N is the total number of atoms. The fact that $\frac{1}{N} \sum_{\mathbf{k}} n_{\mathbf{k}} \neq 1$ can be attributed to quantum fluctuations which are not completely taken into account within this approach [81]. This also explains why deviations of $\frac{1}{N} \sum_{\mathbf{k}} n_{\mathbf{k}}$ from unity are larger for lower dimensions, as visible in Fig. (2.8).

The poles of the Green functions Eqs. (2.90) determine collective excitations of the atomic system. From Eq. (2.97) one notes that the spectral density can be expressed in the form

$$A_{bb^\dagger}(\mathbf{k},\omega) = \sum_i S^i_{bb^\dagger}(\mathbf{k}) \delta(\omega - \omega^i_{bb^\dagger}(\mathbf{k})), \quad (2.111)$$

2.3 Bose-Hubbard Hamiltonian

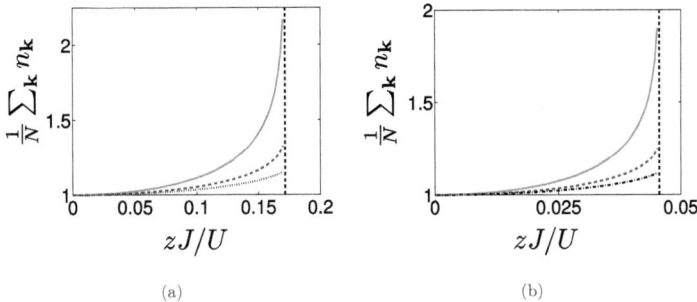

Figure 2.8: (a) Normalization of the momentum distribution as a function of zJ/U for $g = 1$ in one (red solid line), two (blue dashed line) and three (black dotted line) dimensions. (b) same as (a) with $g = 5$. The thin vertical line indicates the phase transition. The pictures are reproduced from Menotti *et al.* [81].

where $\omega^i_{bb^\dagger}(\mathbf{k})$ are the frequencies of the excitation modes and $S^i_{bb^\dagger}(\mathbf{k})$ their respective weights.

We now compare the results obtained from RPA for the spectral density with the results obtained from Bogoliubov theory. The spectral density within Bogoliubov theory reads

$$A^B_{bb^\dagger}(\mathbf{k},\omega) = 2\pi \left(|u_\mathbf{k}|^2 \delta(\omega - \Omega_\mathbf{k}) - |v_\mathbf{k}|^2 \delta(\omega + \Omega_\mathbf{k}) \right), \quad (2.112)$$

where the frequencies and coefficients are defined in Eq. (2.66) and Eq. (2.67). From Eq. (2.112) we see that Bogoliubov theory predicts exactly two modes at frequency $\pm\Omega_\mathbf{p}$. We will see that in the RPA this is not always the case and several modes contribute. In order to evaluate $A_{bb^\dagger}(\mathbf{k},\omega)$ in the superfluid phase we diagonalize numerically the mean-field Hamiltonian Eq. (2.37) and use the result to calculate the coefficients $A_{1,2}(\omega + i\eta)$ from Eq. (2.92). The small but finite imaginary part η is used to determine the positions $\omega^i_{bb^\dagger}(\mathbf{k})$ and strength $S^i_{bb^\dagger}(\mathbf{k})$ of the poles of the spectral density from Eq. (2.93) and Eq. (2.97). The sum rule Eq. (2.99) for the spectral density translates to the condition $\sum_j S^i_{bb^\dagger}(\mathbf{k}) = 2\pi$ for the strength of the poles. We determine the number of modes which we take into account from this condition for the strength $S^i_{bb^\dagger}(\mathbf{k})$ and find agreement better than a few parts in 10^{-4} for all values of zJ/U and all momenta.

In Fig. (2.9a) we show the frequencies $\omega^i_{bb^\dagger}(\mathbf{k})$ of the poles of the spectral density and Fig. (2.9b) displays their respective weights $S^i_{bb^\dagger}(\mathbf{k})$. The frequencies $\omega^i_{bb^\dagger}(\mathbf{k})$ and the weights $S^i_{bb^\dagger}(\mathbf{k})$ are plotted as a function of momentum $\mathbf{k}$ in units of $\frac{\pi}{d_0}$ along the

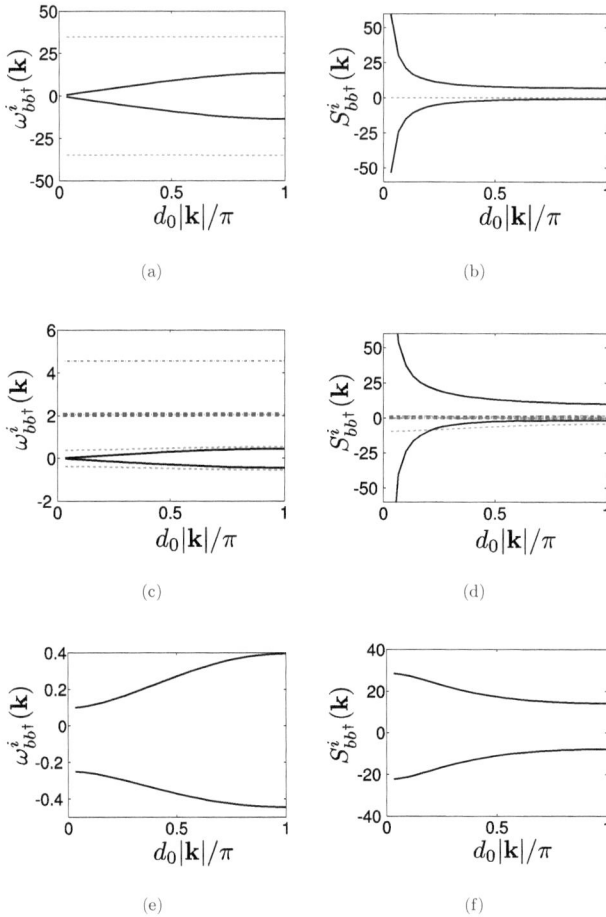

Figure 2.9: (a)(c)(e) Frequencies $w^i_{bb^\dagger}(\mathbf{k})$ and (b,d,f) spectral weights $S^i_{bb^\dagger}(\mathbf{k})$ as defined in Eq. (2.111) as a function of $\mathbf{k}$ in units of $\frac{\pi}{d_0}$ along the $(1,0,0)$ direction. In (a,b) the system is deep in the superfluid state taking $zJ/U = 10$. In (c,d) we assume $zJ/U = 0.2$ such that the system in still in the superfluid state but close to the phase transition. In (e,f) we assume $zJ/U = 0.15$ and the system is in the Mott-insulator phase. For all figures we assume a three dimensional lattice and use $\mu/U = 0.5$.

2.3 Bose-Hubbard Hamiltonian

$(1,0,0)$ direction in the first Brillouin zone, taking $zJ/U = 10$ and $\mu = 0.5$. For these parameters the mean occupation number is given by $\langle b_l^\dagger b_l \rangle \approx 11$ and one obtains for the order parameter $|\Phi|^2 = 10.6$. This gives a condensate fraction of more than 99%. We note that there are 4 modes which contribute to the spectral density with non negligible weight. The black solid lines correspond to the modes predicted from Bogoliubov theory and are the dominant ones as can be seen from their spectral weights. We will refer to these two modes in the following as phonon modes due to the linear dispersion for small momenta. Comparing the phonon modes to the other modes one finds that they contribute more than 98% to the sum rule Eq. (2.99) for all momenta. Comparing the spectral weights $S_{bb^\dagger}^i(\mathbf{k})$ and the frequencies $\omega_{bb^\dagger}^i(\mathbf{k})$ with the analytic results from Bogoliubov theory, agreement is always better than 7%.

Figure (2.9c) displays the frequencies $\omega_{bb^\dagger}^i(\mathbf{k})$ and Fig. (2.9d) their respective weights $S_{bb^\dagger}^i(\mathbf{k})$ as a function of momentum $\mathbf{k}$ in units of $\frac{\pi}{d_0}$ along the $(1,0,0)$ direction in the first Brillouin zone for $zJ/U = 0.2$ with $\mu = 0.5$. For these parameters of the hopping and chemical potential the system is still in the superfluid phase but much closer to the phase transition. The mean occupation number is given by $\langle b_l^\dagger b_l \rangle \approx 1$ and the order parameter has the value $|\Phi|^2 = 0.28$. This leads to a condensate fraction of about 27%. We find 6 modes with non negligible weight where the phonon modes (solid black lines) are the most dominant ones for small momenta. For higher momenta the spectral weight of the phonon modes becomes less dominant with respect to the other modes. For negative frequencies the mode corresponding to the red dashed curve becomes dominant for $k > \pi/4$ as seen from the spectral weights $S_{bb^\dagger}^i(\mathbf{k})$. It describes a collective excitation of the system which is not included in the Bogoliubov treatment. Comparing the frequencies of the phonon modes with the exact analytic expression from Bogoliubov theory one finds large discrepancies of around 30% for all momenta. One can thus conclude, that in the parameter region close to the phase transition Bogoliubov theory is not valid anymore [81].

In Fig. (2.9e) we show frequencies $\omega_{bb^\dagger}^i(\mathbf{k})$ and Fig. (2.9f) their respective weights $S_{bb^\dagger}^i(\mathbf{k})$ as a function of momentum $\mathbf{k}$ in units of $\frac{\pi}{d_0}$ along the $(1,0,0)$ direction in the first Brillouin zone for the system being in the Mott-insulator state as calculated from Eq. (2.106) and Eq. (2.105b). We take $zJ/U = 0.15$ and $\mu/U = 0.5$, such that $\langle b_l^\dagger b_l \rangle = 1$. There are only two modes which contribute to the spectral density and their frequencies correspond to the energy necessary to add or subtract an atom from the system. The difference in excitation energy of the two modes at $k = 0$ corresponds to the gap $\Delta\omega$ of the spectrum in the Mott-insulator phase [81]. It can be calculated from Eq. (2.106) and is given by

$$\Delta\omega = \lim_{|\mathbf{k}|\to 0} (\omega_1 - \omega_2) = \sqrt{(U-zJ)^2 - 4zJUg}\,. \tag{2.113}$$

This frequency gap coincides with the width of the mean-field Mott lobe at the same hopping as can be verified from Eq. (2.45).

Chapter 3

Photonic Band Structure of a Bichromatic Optical Lattice

So far we have discussed the analogs between ultracold atoms in optical lattices and the behaviour of electrons in the crystalline structure in solid state physics. A remarkable feature of optical lattices is that the bulk periodicity is here controlled by engineering the geometry of the propagating beams, which determine the light potentials [14, 15]. Differing from ordinary crystals in condensed matter, the size of the Wigner-Seitz cell in optical lattices is of the order of the light wavelength. One consequence is that the light, coupling with the atomic transitions, is also diffracted by the crystalline structure which the atoms form [82].

It has been observed that the modulation of the atomic density in these systems, and hence of the refractive index, makes optical lattices a photonic bandgap material [83]. Theoretical works studied the photonic bandgap for one-dimensional and three-dimensional atomic structures [84, 83, 85, 86, 87, 88].

In this chapter we study theoretically the photonic properties of biperiodic optical lattices, in a setup similar to the ones realized experimentally in [89, 90, 91]. We focus on a one-dimensional configuration assuming that the atoms are well localized at the lattice minima. The photonic spectra and the probe transmission are evaluated when the optical lattice is in free space and inside a standing wave optical resonator, as a function of the interparticle distance ϱ inside the primitive cell of length a. We first consider the special case of a monochromatic optical lattice as studied in [83, 84] and then extend our study to the case of a bichromatic optical lattice. The transmission spectra of a weak incident probe are evaluated when the atoms are trapped in free space and inside an optical resonator for realistic experimental parameters. We end the chapter with a discussion of the results obtained.

3.1 Theoretical model

The physical system we consider is a one-dimensional periodic distribution of atoms in a light potential with a double primitive cell. We assume a sequence of N atoms of mass m in a standing wave created with lasers along the x direction. We denote the atomic positions along x by x_j, with $j = 1, \ldots, N$. Denoting by a the size of the Wigner-Seitz cell, the positions are given by

$$\begin{aligned} x_j &= \ell a \text{ for } j = 2\ell, \\ &= \ell a + \varrho \text{ for } j = 2\ell + 1, \end{aligned}$$

where $\ell = 0, 1, \ldots, M-1$, and $M = N/2$ is the number of cells (assuming N even for convenience). In the case here discussed, we set the size

$$a = \lambda, \tag{3.1}$$

where λ is the wavelength of the light which interacts with the dipolar transitions of the atoms. Such configuration can be experimentally realized by using a monochromatic standing wave with wavelength λ, to which two laser beams are superposed, such that they are rotated by angles of 60° and 120° with respect to the axis of the lattice, as shown in Fig. (3.1). In the dispersive regime the lasers form a conservative potential for the atoms as discussed in Sec. (2.1). Upon setting the relative phases, the resulting potential for the atoms has the form

$$U(x) \propto \beta^2 \cos^2(kx/2) + \cos^2(kx) \tag{3.2}$$

and the distances between adjacent wells are $d_1 = \lambda(1 - \frac{1}{\pi} \text{acos} \frac{-\beta^2}{4})$ and $d_2 = \frac{\lambda}{\pi} \text{acos} \frac{-\beta^2}{4}$, with $d_1 + d_2 = \lambda$. Another possible realization is found by superposing two laser beams along the x-axis, with a half frequency [92] or with a three fourth frequency respect to the frequency of the main lattice [93]. Upon setting the relative phases, the four-atomic elementary cell has the structure $d_1 - d_1 - d_2 - d_2$ (with $d_1 + d_2 = \lambda$) and the crystal has essentially the same spectral properties than the biatomic one considered in this section. We will also consider the possibility that the atoms composing the Wigner-Seitz cell may have different scattering properties, for instance, they can belong to different species or belong to the same species but are prepared in different hyperfine states. Under the assumption that the frequencies of the two transitions are sufficiently close to allow significant coupling with the same probe. Such lattices could be realized with linearly polarized counterpropagating beams, controlling the angle between the polarization [70].

In developing the theoretical model we will assume that the atoms are well localized at the lattice points, and the size of the atomic wave packet is very small with respect to the laser wavelength[1]. As seen in Sec. 2.3 atom tunneling from site to site is suppressed

---
[1] This regime is also called Lamb-Dicke regime [94].

3.1 Theoretical model

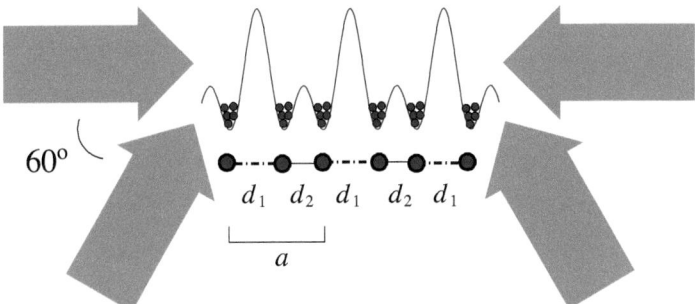

Figure 3.1: A possible optical realization of the double period 1D lattice, considered in this work, can be obtained by using a monochromatic standing wave with wavelength λ, to which two laser beams are superposed, such that they are rotated by angles of 60° and 120° with respect to the axis of the lattice. Here, $d_1 = \varrho$, $d_1 + d_2 = \lambda$ and the size of the Wigner-Seitz cell is $a = \lambda$.

when the system is in the Mott-insulator state. From Eq. (2.24) we can estimate the size ξ of the atomic wave packet and find $\xi \approx 0.09\lambda$ for a monochromatic optical lattice of depth $V_0 = 10E_R$. Thus the assumption of pointlike atoms at fixed positions is well justified when the atoms are deep in the Mott-insulator quantum state of the biperiodic potential [14]. We consider only the coupling of the lattice with modes of the electromagnetic field which propagate along the lattice. This approximation is valid when the atoms are placed, for instance, inside a bad cavity with sufficiently large cooperativity [95] or a hollow-core fiber [96, 97].

3.1.1 Hamiltonian

The Hamiltonian H_e for the system of fixed pointlike particles can be be formally obtained from the Eq. (1.42) by taking the mass of the atoms to infinity $m \to \infty$[2]. In this case the kinetic energy of the atoms vanishes and the atoms are confined at fixed positions, corresponding to the minima of the trapping potential. Thus we find for the total Hamiltonian

$$H = H_e + H_{af} + H_{emf}, \qquad (3.3)$$

---

[2] For finite mass m one has in principle to take into account the finite width ξ of the atomic wavefunctions located at each lattice site. Since we assume that ξ is much smaller than the resonant wavelength coupling to the dipolar transitions of the atoms the corrections to our treatment due to the kinetic energy of the atoms will be negligible.

where
$$H_e = \sum_j \hbar\omega_j \sigma_{ee}(j) \tag{3.4}$$
describes the array of dipoles, with $j = 1, \ldots, N$ labeling the atoms. Hereby the relevant transitions of the atoms at the even (odd) sites, $j = 2\ell$ ($j = 2\ell+1$), have dipole moments $\mathbf{D}^1$ ($\mathbf{D}^2$) and transition frequency ω_1 (ω_2). The interaction between photons and atoms is given in Eq. (1.43). Restricting to modes of the electromagnetic field propagating along the x-direction and omitting the spin indices α and β one gets
$$H_{\text{af}} = \sum_j \sum_\lambda \hbar C_\lambda^j \sigma_{eg}(j) a_\lambda e^{ik_\lambda x_j} + \text{H.c.} \tag{3.5}$$
The coupling C_λ^j is given in Eq. (1.44) and the index j refers here to the position and reflects the fact that the coupling is proportional to the dipole-moment $\mathbf{D}^1$ or $\mathbf{D}^2$ dependent if j is even or odd. The Hamiltonian for the modes of the electromagnetic field is given in Eq. (1.51), where we restrict to modes with wave vectors along the x-axis.

3.1.2 Weak excitation regime

We consider a weak probe field such that the atomic transitions are driven well below saturation, and correspondingly the mean number of photonic excitations inside the system is much smaller than the total number of spins N. In this regime we use the Holstein-Primakoff representation of spin operators [98], and expand all operators at the lowest orders in the powers of bosonic operators b_j,

$$\sigma_{eg}(j) = b_j^\dagger (1 - b_j^\dagger b_j)^{1/2} \simeq b_j^\dagger \left(1 - b_j^\dagger b_j/2\right), \tag{3.6}$$
$$\sigma_{ge}(j) = (1 - b_j^\dagger b_j)^{1/2} b_j \simeq \left(1 - b_j^\dagger b_j/2\right) b_j, \tag{3.7}$$
$$\sigma_j^z = \sigma_{ee}(j) - \sigma_{gg}(j) = -\frac{1}{2} + b_j^\dagger b_j. \tag{3.8}$$

In this representation, the Hamiltonian for the dipoles becomes the sum of N harmonic oscillators,
$$H_e = \sum_j \hbar\omega_j b_j^\dagger b_j, \tag{3.9}$$
where we discarded the constant term. The interaction term reads
$$H_{\text{af}} = H^{(1)} + H^{(3)}, \tag{3.10}$$
with
$$H^{(1)} = \sum_{j,\lambda} \hbar C_\lambda^j b_j^\dagger a_\lambda \, e^{ik_\lambda x_j} + \text{H.c.}, \tag{3.11}$$

3.1 Theoretical model

while $H^{(3)}$ describes the corrections beyond the linear response. We will consider the limit in which we can truncate the expansion and approximate $H_{\text{int}} \approx H^{(1)}$, thereby restricting to the case in which the medium polarization is linear in the electric field amplitude.

3.1.3 Spin waves

Given the periodic structure, it is convenient to describe the dipolar excitations in momentum space. At this purpose, for a sufficiently large crystal we assume Born-von Karman periodic boundary conditions, and consider the spin-wave excitations

$$b_q = \frac{1}{\sqrt{M}} \sum_{\ell=0}^{M-1} b_{2\ell} e^{-i\ell qa}, \tag{3.12}$$

$$d_q = \frac{1}{\sqrt{M}} e^{-iq\varrho} \sum_{\ell=0}^{M-1} b_{2\ell+1} e^{-i\ell qa}, \tag{3.13}$$

with q the wave vector sweeping the first Brillouin zone (BZ). We denote by

$$G_0 = 2\pi/a$$

the elementary vector of the reciprocal lattice, such that the interval of the first BZ is $[-G_0/2, G_0/2]$. Using the relation $\sum_{\ell=0}^{M-1} \exp(i(q-q')\ell a) = M\delta_{qq'}$ where the equality $q = q'$ is defined modulus a vector G of the reciprocal lattice, the Hamiltonian terms transform as

$$H_e = \sum_{q \in BZ} \hbar \left(\omega_1 b_q^\dagger b_q + \omega_2 d_q^\dagger d_q \right), \tag{3.14}$$

$$H^{(1)} = \sum_{G,n} \sum_{q \in BZ} \hbar\sqrt{M} \left(\mathcal{G}_{1,q+G}^{(n)} b_q^\dagger + e^{iG\varrho} \mathcal{G}_{2,q+G}^{(n)} d_q^\dagger \right) a_{q+G}^{(n)} + \text{H.c.}, \tag{3.15}$$

where the index n stands for the different polarizations and the quasi-momentum verifies the relation $k = q + G$. The coupling constants are given by

$$\mathcal{G}_{i,q+G}^{(n)} = \sqrt{\frac{2\pi\omega_{q+G}}{\hbar\mathcal{V}}} \left(\mathbf{D}^i \cdot \hat{\mathbf{e}}_n \right) \tag{3.16}$$

with $i = 1, 2$. In this form, the Hamiltonian can be rewritten as the sum of M Hamiltonian terms, $H = \sum_{q \in BZ} H_q$, where

$$\begin{aligned}H_q &= \hbar\omega_1 b_q^\dagger b_q + \hbar\omega_2 d_q^\dagger d_q + \hbar \sum_{G,n} \omega_{q+G} a_{q+G}^{\dagger(n)} a_{q+G}^{(n)} \\ &+ \hbar \sum_{G,n} \left[\sqrt{M} \left(\mathcal{G}_{1,q+G}^{(n)} b_q^\dagger + e^{iG\varrho} \mathcal{G}_{2,q+G}^{(n)} d_q^\dagger \right) a_{q+G}^{(n)} + \text{H.c.} \right].\end{aligned} \tag{3.17}$$

This separation is only valid in the linear regime since saturation effects, described by $H^{(3)}$, mix the manifolds identified by the Hamiltonian terms H_q.

We will the photonic spectrum of the biperiodic structure assuming that the polarization of the incident light, say ϵ_1, is parallel to the dipole moments $\mathbf{D}^1$ and $\mathbf{D}^2$. Hence, we drop the polarization superscripts where they appear. The photonic band structure is found by solving the Heisenberg equations of motion for each Hamiltonian block H_q.

$$\dot{a}_{q+G} = -i\omega_{q+G}a_{q+G} - i\sqrt{M}\left(\mathcal{G}_{1,q+G}b_q - e^{-iG\varrho}\mathcal{G}_{2,q+G}d_q\right), \quad (3.18a)$$

$$\dot{b}_q = -i\omega_1 b_q - i\sqrt{M}\sum_G \mathcal{G}_{1,q+G}a_{q+G}, \quad (3.18b)$$

$$\dot{d}_q = -i\omega_2 d_q - i\sqrt{M}\sum_G e^{iG\varrho}\mathcal{G}_{2,q+G}a_{q+G}, \quad (3.18c)$$

where $\omega_q = c|q|$.

3.2 Monochromatic optical lattice

We assume for the moment only a single atomic specie in the lattice. Taking $\rho = 0.5a$ the primitive cell of the lattice has length $a/2$ and only contains a single atom. In this case the system is a monochromatic optical lattice as studied in [84, 83]. Since the length of the primitive cell is $a/2$ the Brillouin zone is doubled compared to the bichromatic case. This is shown in Fig. (3.2) in a schematic way. One sees that for a monochromatic optical lattice $\rho = 0.5a$ the Brillouin zone ranges from $[-\frac{2\pi}{a} \ldots \frac{2\pi}{a}]$ such that the resonant photonic modes are at the edges of the Brillouin zone. Due to the interaction between the photons and the atomic dipoles a bandgap opens up at $q = \pm\frac{2\pi}{a}$ [84, 83]. This is indicated by the solid blue line which shows the polariton branches in the case of a monochromatic optical lattice. In the monochromatic case Eq. (3.18) reduces to

$$\dot{a}_{q+2G} = -i\omega_{q+2G}a_{q+2G} - i\sqrt{2M}\mathcal{G}_{1,q+2G}b_q, \quad (3.19)$$

$$\dot{b}_q = -i\omega_1 b_q - i\sqrt{2M}\sum_G \mathcal{G}_{1,q+2G}a_{q+2G}. \quad (3.20)$$

The replacement $M \to 2M$ comes from the fact that we have effectively doubled the number of cells, since the primitive cell has now length $a/2$. The spin wave at wave vector q couples in principle with all photonic modes at wave vectors $q + 2G$. Nevertheless, only the coupling of the atomic transition with the quasi-resonant modes at wave vectors $Q = \pm G_0$ is significant. Taking into account only the relevant coupling, we can solve

3.2 Monochromatic optical lattice

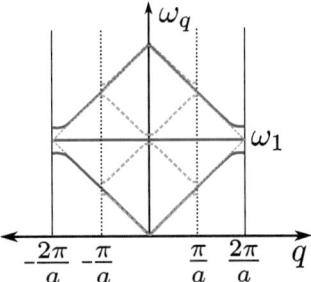

Figure 3.2: Schematic drawing of the polariton branches in the case of a monochromatic optical lattice $\rho = 0.5a$ and in the case of a bichromatic optical lattice. The solid black line indicates the Brillouin boundary if $\rho = 0.5a$ and the black dotted line indicates the edge of the Brillouin zone in the bichromatic case $\rho \neq 0.5a$. The red dashed lines indicate the unperturbed photonic and atomic energies. The blue solid line is a sketch of the polariton energies in the monochromatic case $\rho = 0.5$ and the green dot-dashed line is a sketch op the polariton energies in the bichromatic case $\rho \neq 0.5a$.

analytically the eigenvalue problem around $q \simeq \frac{2\pi}{a}$ at the edge of the BZ to get [84]

$$\nu_0 = \omega_Q, \tag{3.21a}$$

$$\nu_\pm = \frac{\omega_Q + \omega_1}{2} \pm \sqrt{\left(\frac{\omega_Q - \omega_1}{2}\right)^2 + 4M\mathcal{G}_{1,Q}^2}. \tag{3.21b}$$

The eigenfrequency ν_0 in Eq. (3.21) corresponds to a polariton mode which is created by the operator $(a_Q^\dagger - a_{-Q}^\dagger)/\sqrt{2}$. It corresponds to the standing light wave that creates the optical lattice. In the resonant case $\omega_1 = \omega_Q$ these eigenfrequencies determine the edge of the photonic bandgap. We note that the bandgap size is independent on the number of cells M, and thus it is constant in the thermodynamic limit: in fact the quantization volume $\mathcal{V} \propto 1/\sqrt{M}$ (in 1D) gives that $\mathcal{G}_{1,Q} \propto 1/\sqrt{M}$, so that the dependence on M in Eq. (3.21) cancels out.

In Fig. (3.3) we plot the polariton dispersion relation for a monochromatic 1D optical lattice for exact resonance $\omega_1 = \omega_Q$ and in the detuned case $\omega_1 = \omega_Q - 530\gamma$ around $q = \frac{2\pi}{a}$. We note that the bandgap is largest in the resonant case. In the case of resonance one of the polariton branches is hidden since it lies exactly at the unperturbed atomic energy and corresponds to the solution ν_0 in Eq. (3.21). In the detuned case we note that two bandgaps arise, a polaritonic one at the unperturbed atomic energy and a photonic one at the edge of the Brillouin zone.

We note that Eq. (3.21) is equivalent with the expression for the vacuum Rabi splitting of a two level atom inside a cavity with quantization volume $\mathcal{V}_c = \frac{\mathcal{V}}{2M} = a/2$

Photonic Band Structure of a Bichromatic Optical Lattice

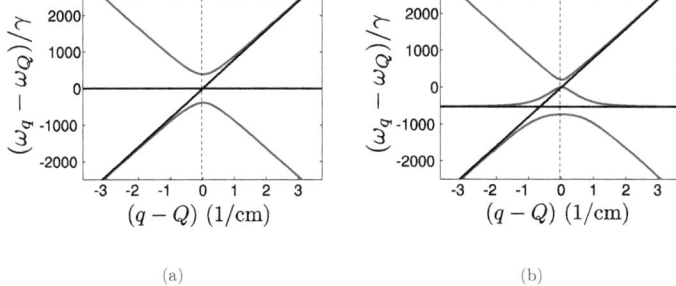

Figure 3.3: Polariton dispersion for a 1D monochromatic lattice obtained by summing over 40 BZ, when the atoms are trapped in a hollow fiber, whose fundamental mode is Gaussian with waist $w = 5\mu m$ [97]. We consider the D2-line of ^{85}Rb atoms, with $\lambda = 780$ nm and $\gamma = 2\pi \times 6$ MHz. The solid black line is plotted for comparison, and corresponds to the case in which atoms and field are not coupled ($\mathcal{G}_j = 0$) and the dashed black line marks the edge of the Brillouin zone. (a) resonant case where $\omega_1 = \omega_Q$ (b) detuned case where $\omega_1 = \omega_Q - 530\gamma$

[49]. Indeed one could interpret the vacuum Rabi splitting of an atom in a cavity as a bandgap at the atomic resonance due to the monochromatic optical lattice made by the atom inside the cavity and all its images [99]. This analogy is only valid in the linear regime which corresponds in the single atom case to a single excitation.

3.3 Biperiodic optical lattice

Let us now consider the photonic properties of a bichromatic optical lattice. Displacing every second atom such that $\rho \neq 0.5a$ we change the symmetry of the system and the size of the primitive cell changes to a. In this case the new Brillouin zone ranges from $[-\frac{\pi}{a} \ldots \frac{\pi}{a}]$. The resonant photonic modes are now at the center of the new Brillouin zone and the photonic gap lies at $q = 0$, as can be seen schematically in Fig. (3.2). The polariton branches are folded into the new Brillouin zone as is indicated by the dot-dashed green lines in Fig. (3.2) and we see that a new pair of bandgaps opens up at $q = \pm\frac{\pi}{a}$. However the photonic modes at $q = \pm\frac{\pi}{a}$ are very far detuned from the atomic resonance such that the width of these gaps are negligible. Restricting to the quasi resonant modes with wavevectors $q = \pm G_0$ and assuming $\omega_1 = \omega_2$ one finds four eigenfrequencies from Eq. (3.18),

$$\nu_{j,\pm} = \frac{\omega_Q + \omega_1}{2} \pm \sqrt{\left(\frac{\omega_Q - \omega_1}{2}\right)^2 + M\mathcal{G}^2 \left(1 - (-1)^j \sqrt{1 - \left(\frac{2|\mathcal{G}_{1,Q}\mathcal{G}_{2,Q}|}{\mathcal{G}^2}\right)^2} \sin^2 G_0 \varrho\right)}. \tag{3.22}$$

Here $\mathcal{G} = \sqrt{|\mathcal{G}_{1,Q}|^2 + |\mathcal{G}_{2,Q}|^2}$ and $j = 1, 2$. The eigenfrequencies determine the edges of two photonic bandgaps, one at the frequencies between $\nu_{1,-}$ and $\nu_{2,-}$ and the second between $\nu_{2,+}$ and $\nu_{1,+}$. We note that the bandgap size depends on the interparticle distance ϱ inside the Wigner-Seitz cell but is independent on the number of cells M. For $\varrho \to 0$, in the limit of the monoperiodic array with the same primitive cell as the bichromatic optical lattice, one finds a single bandgap with size $\Delta\omega = \nu_{1,+} - \nu_{1,-}$. This coincides with the bandgap at $\rho = 0.5a$ as discussed in Sec. (3.2) and shows that in the linear regime the two situations are identical.

For $\varrho > 0$ ($\varrho < a$) $\Delta\omega$ is reduced: a frequency window opens inside the gap, where light is transmitted, and whose width is given by $\Delta\omega_\varrho = \nu_{2,+} - \nu_{2,-}$. The size of the two bandgaps is minimum at $\varrho = a/4$, and it vanishes at this point when $\mathcal{G}_{1,Q} = \mathcal{G}_{2,Q}$. In this specific case, hence, the lattice becomes completely transparent. This is simply understood, considering that the bandgap results from an interference effect due to multiple scattering by all atomic planes, and it hence depends on the phase relations between the fields scattered by each plane. For this specific configuration, where $\varrho = \lambda/4, 3\lambda/4$, the phase accumulated due to scattering of the first atom of the cell cancels out with the phase due to scattering by the second atom. As a result, the total phase accumulated from scattering with the two atoms of the cell is zero, and the medium hence behaves as it were completely transparent.

These analytical results, obtained in a specific parameter regimes, are confirmed by the results of the numerical spectra, which are evaluated from Eqs. (3.18) by summing over 40 BZs. The photonic spectra are shown in Figs. 3.4-3.5(a), where the polariton dispersion relation is reported around $q \simeq 0$ for $\varrho = 0, 0.2a, 0.4a$ for several values of ω_1, setting $\omega_1 = \omega_2$. The size of the bandgaps $\Delta \omega_+ = \nu_{2,+} - \nu_{1,+}$ and $\Delta \omega_- = \nu_{2,-} - \nu_{1,-}$ as a function of ϱ are displayed in Figs. 3.4-3.5(b), showing that the size of the gap is controlled by ϱ, and it vanishes at. Figures 3.6 and 3.7 display the photonic spectra

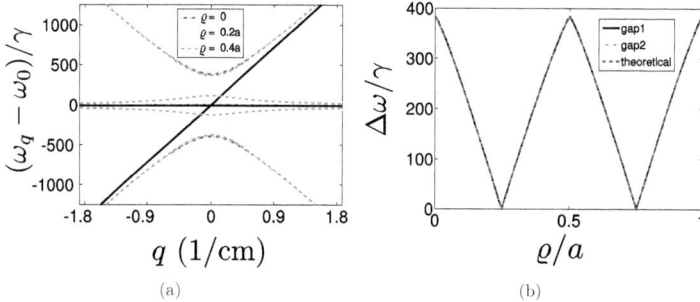

Figure 3.4: (a) Polariton dispersion for a 1D bichromatic lattice for different values of ϱ. The curves are evaluated for $\omega_1 = \omega_2 = \omega_0 - 10\gamma$ and all other parameters are chosen as in Fig. (3.3). (b) Photonic bandgap (numbered from lower to higher frequency) as a function of ϱ/a, compared with the analytical prediction obtained from Eq. (3.22).

when the atoms composing the Wigner-Seitz cell are of different species, in the case in which both interact with the probe but the resonance frequency of the respective dipolar transition is different. Figure 3.6 displays the photonic spectrum for the specific case in which one atomic transition is quasi-resonant, while the second is far detuned. In this case three photonic bandgaps appear, which vary largely as a function of ϱ, and in such a way that while one is minimum, the other two are maximum, and vice versa. Figure 3.7 displays the case in which the two atoms composing the cell are far detuned from the probe, with detunings of opposite signs. The spectrum is also characterized by three bandgaps.

In this treatment we neglected atomic absorption, however the evaluated bandgaps are significantly larger than the linewidth γ, and hence they can be experimentally observed.

3.3 Biperiodic optical lattice

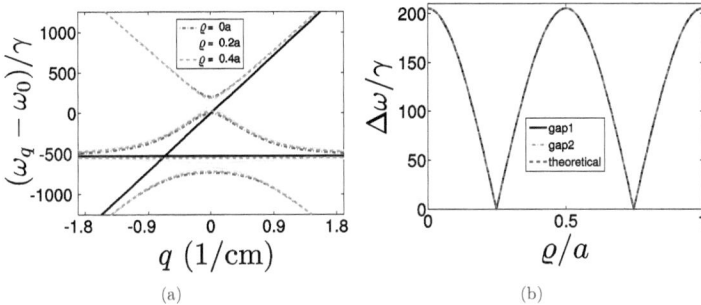

Figure 3.5: Same as in Fig. (3.4) but for $\omega_1 = \omega_2 = \omega_0 - 530\gamma$.

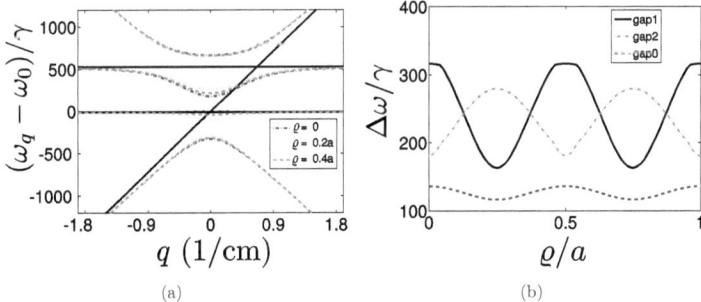

Figure 3.6: Same as in Fig. 3.4 but for $\omega_1 = \omega_0 - 10\gamma$ and $\omega_2 = \omega_0 + 530\gamma$.

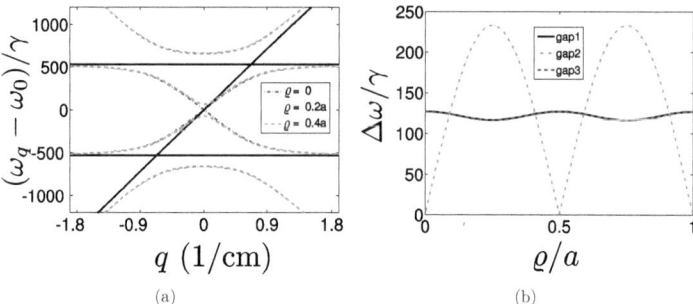

Figure 3.7: Same as in Fig. (3.4) but for $\omega_1 = \omega_0 - 530\gamma$ and $\omega_2 = \omega_0 + 530\gamma$. Note that gap 1 and 3 overlap.

Figure 3.8: Schematic sketch of a bichromatic lattice inside a optical resonator. The cavity is pumped by a coherent light beam with amplitude η and frequency ω_p and the transmitted light intensity $I(\omega_p)$ is measured as a function of ω_p.

3.4 Bichromatic lattice inside a single-mode cavity

The observation of sufficiently large photonic bandgaps in free space requires a large number of lattice planes in a well controlled periodic structure, which is experimentally challenging. Nevertheless, observable effects can be found in small systems when coupling the atomic transition, for instance, to the modes of a fiber [97] or of an optical resonator [100, 101, 102].

Let us assume that the dipolar transition of the atoms couples strongly with the single mode of a standing-wave optical resonator, which probes the system as shown in Fig. (3.8). The Hamiltonian for the dynamics inside the cavity is now $H' = H_e + H_c + H'_{af}$, where H_e is given in Eq. (3.4),

$$H_c = \hbar\omega_c a^\dagger a \qquad (3.23)$$

describes the cavity field at frequency ω_c, with a, $a^\dagger$ annihilation and creation operators

3.4 Bichromatic lattice inside a single-mode cavity

of a cavity photon, and

$$H'_{\text{af}} = \hbar \sum_j g_j \cos(kx_j + \varphi)\sigma_{eg}(j)a + \text{H.c} \qquad (3.24)$$

is the Jaynes-Cummings Hamiltonian, with k the cavity mode wave vector and g_j the coupling strength, which depends on the dipolar moment $\mathbf{D}^j$ of the atom at position x_j. As a function of the cavity parameters, $g_j = \sqrt{\varsigma/(4\pi A)}\sqrt{\gamma\delta\omega}$, with ς the scattering cross section in free space, $A = \pi w_c^2/4$ with w_c the cavity mode waist, and $\delta\omega = 2\pi c/L$ the free spectral range, with L the cavity length [95]. The phase φ accounts for the dephasing between the cavity mode lattice and the atomic lattice.

The system is probed by an external weak coherent pump at intensity η and frequency ω_p, which is coupled to the resonator. In this limit, we make the Holstein-Primakoff transformation and keep only the linear term. The resulting Heisenberg-Langevin equations of motion for cavity mode and spin wave operators read [103]

$$\begin{aligned}
\dot{a} &= -\mathrm{i}\delta_c a - \kappa a + \eta + \sqrt{2\kappa}\zeta(t) - \mathrm{i}\frac{\sqrt{M}}{2} \\
&\quad \times \sum_{Q \in BZ, Q = k-G} \left[g_1 \left(\mathrm{e}^{-\mathrm{i}\varphi}b_Q + \mathrm{e}^{\mathrm{i}\varphi}b_{-Q} \right), \right. \\
&\quad \left. + g_2 \left(\mathrm{e}^{-\mathrm{i}(k\varrho+\varphi)}d_Q + \mathrm{e}^{\mathrm{i}(k\varrho+\varphi)}d_{-Q} \right) \right] \qquad (3.25\text{a}) \\
\dot{b}_{\pm Q} &= -\left(\mathrm{i}\delta_1 + \frac{\gamma_1}{2}\right)b_{\pm Q} - \frac{\mathrm{i}}{2}\sqrt{M}g_1 \mathrm{e}^{\pm \mathrm{i}\varphi}a + \sqrt{\gamma_1}\mathcal{B}_{1,Q}, \qquad (3.25\text{b}) \\
\dot{d}_{\pm Q} &= -\left(\mathrm{i}\delta_2 + \frac{\gamma_2}{2}\right)d_{\pm Q} - \frac{\mathrm{i}}{2}\sqrt{M}g_2 \mathrm{e}^{\pm \mathrm{i}(k\varrho+\varphi)}a + \sqrt{\gamma_2}\mathcal{B}_{2,Q}, \qquad (3.25\text{c})
\end{aligned}$$

where $\delta_c = \omega_c - \omega_p$, $\delta_j = \omega_j - \omega_p$ and κ is the cavity linewidth. The noise operators $\zeta(t)$, $\mathcal{B}_{j,Q}$, have zero mean value and satisfy the relation $\langle \zeta(t)\zeta(t')^\dagger \rangle = \langle \mathcal{B}_{j,Q}(t)\mathcal{B}_{j,Q}(t')^\dagger \rangle = \delta(t-t')$ (we assume the electromagnetic field in the vacuum).

Equations (3.25a)-(3.25c) describe the coupling between the cavity mode, at momentum k, and the spin waves at quasi-momentum Q (inside the first Brillouin zone), such that $Q + G = k$ where G is a vector of the reciprocal lattice. We identify two relevant cases, when (i) $k \neq \mathcal{N}\pi/a$, and (ii) $k = \mathcal{N}\pi/a$, where $\mathcal{N}$ is an integer.

For $k \neq \mathcal{N}\pi/a$ the system, composed by cavity potential and bichromatic lattice, is not periodic. The eigenfrequencies of the homogeneous equations can be simply found for the case $\delta_1 = \delta_2 = \Delta$, $\gamma_1 = \gamma_2 = \gamma$, and read

$$\nu_0 = \Delta - \mathrm{i}\gamma/2, \qquad (3.26)$$

$$\nu_\pm = \frac{\delta_c + \Delta - \mathrm{i}(\kappa + \gamma/2)}{2} \pm \sqrt{\frac{1}{4}\left(\delta_c - \Delta - \mathrm{i}\kappa + \mathrm{i}\frac{\gamma}{2}\right)^2 + M\mathcal{R}}, \qquad (3.27)$$

where
$$\mathcal{R} = (g_1^2 + g_2^2)/2. \tag{3.28}$$

Here, the real part gives the position of the resonances, while the imaginary part gives the corresponding linewidth. The eigenmodes at frequency ν_0 are pure spin waves, and hence correspond to collective dipolar excitations which are decoupled from the cavity field. The eigenmodes at frequency $\nu_\pm$ are polariton excitations. We remark that the frequencies $\nu_\pm$ do not depend on ϱ.

For $k = \mathcal{N}\frac{\pi}{a}$, the system, composed by cavity potential and bichromatic lattice, is periodic. The cavity mode couples to the spin waves $Q = 0$ or π/a, depending on whether $\mathcal{N}$ is even or odd, respectively. The eigenfrequencies of the homogeneous equations (for the specific case $\delta_1 = \delta_2 = \Delta$, $\gamma_1 = \gamma_2 = \gamma$) have the same form as in Eqs. (3.26-3.27), whereby now in Eq. (3.27) the coefficient $\mathcal{R}$ reads

$$\mathcal{R} = g_1^2 \cos^2\varphi + g_2^2 \cos^2(k\varrho + \varphi). \tag{3.29}$$

We note that the eigenfrequencies in this case explicitly depend on ϱ. This result is also found when the bichromatic lattice is replaced by one single cell, by rescaling the coupling strength of each atom inside the cell as $g_{\text{eff},j} = \sqrt{M}g_j$.

We now discuss the intensity of the field at the cavity output as a function of the probe frequency ω_p for various parameters, by solving Eqs. (3.25a)-(3.25c) in the steady state. The quantity we study is the number of photons per unit time $I(\omega_p) = \langle a_{\text{out}}^\dagger a_{\text{out}} \rangle$, where a_{out} is the field at the cavity output, $a_{\text{out}} = \sqrt{2\kappa}a - \zeta$, and the average is taken over the state of the system and the vacuum state of the e.m.-field outside the resonator [103]. Hence, we find

$$I(\omega_p) = 2\kappa \langle a^\dagger a \rangle = \frac{2\kappa\eta^2(\Delta^2 + \gamma^2/4)}{(\kappa\gamma/2 - \delta_c\Delta + M\mathcal{R})^2 + (\Delta\kappa + \delta_c\gamma/2)^2}, \tag{3.30}$$

where we have used the steady state solution of Eq. (3.25a). In the strong coupling regime, when the cooperativity $\mathcal{C} \sim M\mathcal{R}/2\kappa\gamma \gg 1$ [95], the intensity $I(\omega_p)$ at the cavity output exhibits two well defined maxima at the frequencies

$$\omega_p^0 = \frac{\omega_c + \omega_a}{2} \pm \sqrt{\left(\frac{\omega_c - \omega_a}{2}\right)^2 + M\mathcal{R}}, \tag{3.31}$$

which correspond to the vacuum Rabi splitting for this system [95]. This can also be seen in Figs. 3.9 and 3.10. It is interesting to note that for $\Delta = 0$ and large cooperativity the cavity output field goes to zero as $1/\mathcal{C}^2$. This is an interference effect, where the atomic polarization inside the cavity form a field equal and opposite to the driving pump, such that the cavity field is effectively empty. Energy is in this case dissipated by the atoms. This behaviour has been first predicted in [104] under the name "cavity induced transparency".

3.4 Bichromatic lattice inside a single-mode cavity

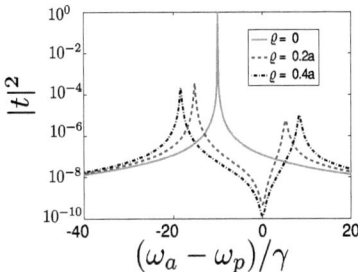

Figure 3.9: (Color online) The absolute square of the transmission coefficient as a function of $(\omega_a - \omega_p)$ (in units of γ) is plotted for a lattice of ^{85}Rb. The relevant transition is the D2-line with $\lambda = 780$nm and $\gamma = 2\pi \times 6$MHz. The atoms are trapped by a laser beam detuned to the blue of resonance by $10\,\gamma$ which is in resonance with the cavity. We assume $M = 100$ primitive cells with a mean occupation number $\bar{n} = 3000$ for each well and take $\varphi = \pi/2$. The cavity decay is given by $\kappa = 2\pi \times 21$kHz. Note that for $\varrho = 0$ the atoms are trapped at the nodes of the resonator and hence do not interact with the cavity field.

We evaluate the cavity transmission spectrum using the parameters of the setup in [105], and consider ^{85}Rb atoms inside a resonator with resonance frequency $\omega_c = \omega_a + 10\gamma$, loss rate $\kappa = 2\pi \times 21$ KHz, beam waist $w = 130$ μm, and an average occupation per site $\bar{n} = 3000$, that corresponds to an areal density $n_s \simeq 5.7 \cdot 10^{-2} \mu\text{m}^{-2}$. The transmission spectrum is calculated from Eq. (3.30) by taking $|t|^2 = I(\omega_p)/I_{in}$ where $I_{in} = \frac{2\eta^2}{\kappa}$. The positions of the peaks correspond to those predicted by Eq. (3.31) by taking into account multiple occupancy of the lattice sites and rescaling the coupling strengths as $g_j \to \sqrt{\bar{n}} g_j$. Figure 3.9 displays the squared transmission as a function of the probe frequency for the case $\varphi = \pi/2$ and the values $\varrho = 0, 0.2a, 0.4a$. Note that the case $\varrho = 0$ corresponds to all atoms at the antinodes of the resonator, and it is hence equivalent to the situation in which the cavity is empty. Figure 3.10 displays the transmission spectrum is the optical lattice trapping the atom is shifted so that $\varphi = 0$, showing that the form changes substantially. By varying φ, hence, information on the interparticle distance in the Wigner-Seitz cell can be gained. The minimum at $\omega_p = \omega_a$, corresponding to the cavity induced transparency behaviour, is here visible.

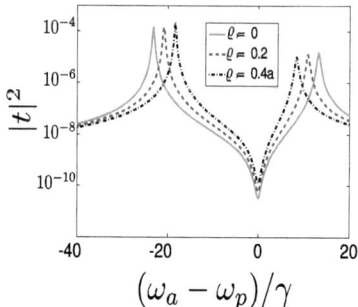

Figure 3.10: Same as Fig. (3.9), but for $\varphi = 0$.

3.5 Discussion

The photonic properties of biperiodic optical lattices are critically determined by the interparticle distance ϱ inside the primitive Wigner-Seitz cell. We have derived a model describing light propagation for a weak probe, and its response to probe propagation in free space and inside of a cavity. In the case that $\varrho = 0.5a$ the system reduces to a monochromatic optical lattice and our results reduce to the ones found in [84]. In the biperiodic case we found that, depending on ϱ, in free space the system may or may not exhibit photonic bandgaps about the atomic frequency. This is a peculiar property, which makes the biperiodic crystal different from the monoperiodic one, always exhibiting a bandgap at the atomic frequency. In case there are photonic bandgaps around this value, they occur in two or more ranges of frequencies. For a finite crystal, relevant effects can be observed when the atoms are confined inside an optical resonator. Here, the interparticle distance ϱ inside the primitive Wigner-Seitz cell determines the properties of the transmission spectrum of a probe at the cavity output.

Our study is based on a full quantum model for the light. In this chapter we have focused on the intensity of the transmitted and reflected light. It would be interesting to study higher order coherence of the scattered light. On the basis of studies made with two atoms inside a cavity [106], one expects that, when considering saturation effects, the biperiodic optical lattice can act as nonlinear-optical medium, whose properties may be controlled by the interparticle distance ϱ.

In the present chapter we neglected the motion of the atoms taking them to be at fixed positions in space. An interesting question is how the photonic band structure

3.5 Discussion

changes if one takes into account the atomic motion and also the statistics of the atoms. A complete diagonalization of the Hamiltonian in order to find the photonic band structure as has been done in Eq. (3.22) will be difficult since the motion of the atoms will break the discrete symmetry of the atomic lattice.

On the other hand a photonic bandgap in 1 dimension can also be seen as Bragg scattering of the incident light since it arises from a destructive interference from the light scattered at all the lattice planes [71]. In the following chapter we will consider the problem of Bragg scattering of ultracold atoms in an optical lattice. We take into account the motion of the atoms and evaluate the scattered light signal taking into account the many body state of the atoms along the Mott-insulator superfluid phase transition.

Chapter 4

Light Scattering from Ultracold Atoms in an Optical Lattice

Bragg scattering in condensed matter is a powerful method for gaining information over the structural properties of crystalline solids. Usually, one employs thermal neutron beams, whose thermal wavelength is of the order of the interparticle distance inside the crystal. While elastic scattering allows one to measure the reciprocal lattice primitive cell, inelastic scattering gives information about the phonon spectrum and anharmonicities [71]. In atomic systems, Bragg scattering has been applied for demonstrating long–range order in structures of cold ions in traps [107] and neutral atoms in optical lattices [108, 109, 82, 110, 100, 101, 102].

Moreover it has proven to be a precise tool for the measurement of the elementary excitations of trapped Bose-Einstein condensate [97, 18] and strongly-correlated atoms in optical lattices [19]. The spectra of the scattered photons, moreover, provide information on the details of atom-photon interactions. Studies on opto-mechanical systems, for instance, showed that the Stokes and anti-Stokes components of the scattered light may exhibit entanglement, which emerges from and is mediated by the interaction with the quantum vibrational modes of the scattering system [111]. Such correlations are endorsed by quantum interference in the processes leading to photon scattering, which is mainly visible in the height of the spectral peaks as a function of the emission angle [112], and can be an important resource for quantum networks [113, 114].

In this chapter we investigate the opto-mechanical properties of strongly-correlated atoms in optical lattices. We study light scattering by ultracold bosonic atoms in an optical lattice, in the setup sketched in Fig. 4.1. We use a full quantum description of the photonic and atomic fields, for a range of optical lattice depths which covers the superfluid to Mott-insulator transition. By starting from the general Hamiltonian Eq. (2.2), we carry out the tight-binding and single-band approximations, and we determine the scattering cross section of photons in the linear response regime.

Extending previous works [32], we systematically take into account the finite tunneling rate in evaluating the scattering cross section for parameters sweeping along the phase transition Mott-insulator to superfluid state. We focus on a small lattice of 7 sites, and solve numerically the Bose-Hubbard model for this system. In order to get insight into the numerical results, we also develop an analytical theory, which extends the theory presented in [21] by including the hopping induced by photon recoil. The interference between the finite tunneling rate and the photon-induced hopping is visible in the height of the Stokes peaks as a function of the emission angle and can be revealed experimentally. We end the chapter with a discussion of the results obtained and comparing them to previous works.

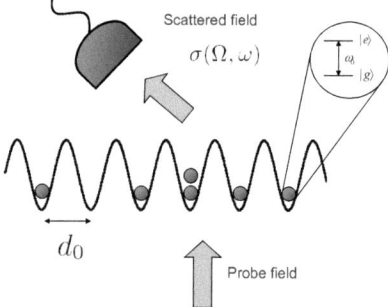

Figure 4.1: Light scattering by atoms trapped in a one-dimensional optical lattice with lattice constant d_0. The atoms are probed by a laser beam, with wave vector $\mathbf{k_L}$ and frequency ω_L, which couples to the atomic dipole transition at frequency ω_0 with ground state $|g\rangle$ and excited state $|e\rangle$ (see inset). The spectrum of the scattered light is measured at a detector as a function of the angle of emission. In experiments, one can also use a second laser beam, into which the photon is emitted with high probability, hence implementing stimulated Bragg scattering [97].

4.1 The Model

The scattering system we consider is composed by N identical bosonic atoms of mass m in a periodic potential, as shown in Fig. 4.1. The relevant internal degrees of freedom of the atoms are the electronic ground state $|g\rangle$ and an excited state $|e\rangle$ that form a dipolar transition with dipole moment $\mathbf{D}^{e,g}$ at the optical frequency ω_0, which couples to a weak laser probe. At room temperature and equilibrium, the atoms are in the electronic ground state and the state of the optical modes of the electromagnetic field can be approximated with the vacuum $|0\rangle$. We now assume that the laser, at frequency

4.1 The Model

ω_L and wave vector $\mathbf{k_L}$, couples to the atomic dipole transition. The laser field is described by a coherent state of the corresponding electromagnetic field mode with amplitude α_L, such that the mean number of photons is given by $|\alpha_L|^2$. We are in the regime in which the atom-laser coupling is sufficiently weak, corresponding to the condition $|C_L^{e,g}\alpha_L| \ll |\omega_0 - \omega_L|$ such that the occupation of the excited state is small and we can use the effective Hamiltonian Eq. (2.1) (discarding the spin index α since we only have a single electronic ground state) to describe the system. The conservative potential $V(\mathbf{r})$ for the ground state atoms is assumed to be periodic along the x-direction and is given in Eq.(2.26).

We assume that the atomic wave functions are well localized at the lattice minima, such that the tight-binding approximation can be applied. Using the expansion Eq. (2.27) for the atomic field operators and following the procedure of Sec(2.3) we find for the total Hamiltonian

$$H = H'_g + H'_{af} + H_{emf}, \qquad (4.1)$$

where

$$H'_g = -J\sum_l b_l^\dagger(b_{l-1} + b_{l+1}) + \frac{U}{2}\sum_l n_l(n_l - 1) - \mu\sum_l n_l \qquad (4.2)$$

is the 1D Bose-Hubbard Hamiltonian Eq. (2.28) as discussed in Sec(2.3) and describes the unperturbed atomic dynamics. The coefficients for the hopping term J and the on-site interaction strength U are given in Eq. (2.29). The atom-field interaction is obtained by using the expansion Eq. (2.27) in the effective atom light interaction Eq. (2.2) and is given by

$$H'_{af} = \sum_{\lambda,\lambda'} \frac{\hbar C_\lambda^{e,g} C_{\lambda'}^{e,g}}{\omega_{\lambda'} - \omega_0} a_\lambda^\dagger a_{\lambda'} \mathcal{T}(\mathbf{q}), \qquad (4.3)$$

with $\mathbf{q} = \mathbf{k}_{\lambda'} - \mathbf{k}_\lambda$. The atomic part in the atom-field Hamiltonian is given by

$$\mathcal{T}(\mathbf{q}) = \sum_l e^{iq_x l d_0}\left[J_0(\mathbf{q})n_l + J_1(\mathbf{q})\left(b_l^\dagger b_{l+1} + b_{l+1}^\dagger b_l\right)\right] \qquad (4.4)$$

and consists of a photon-dependent energy shift, weighted by the coefficient

$$J_0(\mathbf{q}) = e^{-\frac{1}{4}(q_y^2 + q_z^2)\xi_r^2}\int dx e^{iq_x x} w_0(x)^2, \qquad (4.5)$$

with $w_0(x)$ beeing the wannier function centered around 0 and a hopping term with coefficient

$$J_1(\mathbf{q}) = e^{-\frac{1}{4}(q_y^2 + q_z^2)\xi_r^2}\int dx w_0(x) e^{iq_x x} w_0(x - d_0). \qquad (4.6)$$

The latter describes light-assisted tunneling due to the mechanical effects of photon scattering. This latter term has been neglected in previous theoretical treatments [21,

32, 33]. Its effect has been investigated in Ref. [34, 47] for light scattering by ultracold atoms in a double well potential, showing that the mechanical effect of light can interfere with ordinary tunneling between the wells, generating observable effects in the first–order coherence properties of the scattered light. We hence expect that it will give rise to observable effects in the Bragg signal by ultracold atoms in optical lattices. In the following we will assume that the interaction between photons and atoms is essentially Hamiltonian, and hence fully determined by the Schrödinger equation governed by Eq. (2.1). This is valid in the regime which we consider in this article, namely, when the detuning of the light $|\omega_0 - \omega_L| \gg \gamma$, with γ the linewidth of the excited state. In order to study Bragg scattering of laser photons by atoms in the one-dimensional periodic array given by potential (2.26) we will evaluate the differential scattering cross section for coherent scattering. Assuming that $|\alpha_L| \ll 1$, so that the atoms absorb at most one photon from the laser at a time, the differential scattering cross section is found from the rate of scattering one laser photon into the mode λ. In particular, the scattering rate reads

$$\Gamma_{\lambda_L \to \lambda} = \frac{2\pi}{\hbar^2} \sum_f \left| \langle 1_\lambda, f | H'_{\text{int}} | 1_{\lambda_L}, i \rangle \right|^2 \delta^{(t/2)}(\omega_L - \omega_\lambda + (E_i - E_f)/\hbar) ,$$

where we denoted by $|1_\lambda\rangle = a_\lambda^\dagger |0\rangle$ the state of the electromagnetic field with one photon in mode λ, and by $|i\rangle$ and $|f\rangle$ the states of the atoms before and after the scattering, respectively, which are eigenstates of Hamiltonian H'_g at energies E_i and E_f. The function

$$\delta^{(t)}(\omega) = \frac{\sin(\omega t)}{\pi \omega} \tag{4.7}$$

is the diffraction function, giving energy conservation for infinite interaction times, $\lim_{t \to \infty} \delta^{(t)}(\omega) = \delta(\omega)$ [69]. Equation (4.7) shows clearly that the scattering rate depends on the state of the atoms before and after the scattering event. In the case that the atoms are in the superfluid regime, it is convenient to rewrite operator $\mathcal{T}(\mathbf{q})$ in Eq. (4.4) using the decomposition of operator b_l in Eq. (2.34),

$$\mathcal{T}_{\text{SF}}(\mathbf{q}) = \mathcal{T}_{\text{SF}}^{(0)}(\mathbf{q}) + \mathcal{T}_{\text{SF}}^{(1)}(\mathbf{q}) + \mathcal{T}_{\text{SF}}^{(2)}(\mathbf{q}) , \tag{4.8}$$

The first term on the right-hand side of the equation describes radiation coupling with the condensate and reads

$$\mathcal{T}_{\text{SF}}^{(0)}(\mathbf{q}) = g \left(J_0(\mathbf{q}) + 2J_1(\mathbf{q}) \right) \sum_l e^{iq_x l d_0} , \tag{4.9}$$

while the other terms give radiation coupling with the Bogoliubov excitations, and take the form

$$\mathcal{T}_{\text{SF}}^{(1)}(\mathbf{q}) = \sqrt{g}\sum_l e^{iq_x ld_0}\left(J_0(\mathbf{q})(\beta_l + \beta_l^\dagger) + J_1(\mathbf{q})(\beta_l^\dagger + \beta_{l+1} + \beta_{l+1}^\dagger + \beta_l)\right), \quad (4.10)$$

$$\mathcal{T}_{\text{SF}}^{(2)}(\mathbf{q}) = \sum_l e^{iq_x ld_0}\left(J_0(\mathbf{q})\beta_l^\dagger \beta_l + J_1(\mathbf{q})(\beta_l^\dagger \beta_{l+1} + \beta_{l+1}^\dagger \beta_l)\right), \quad (4.11)$$

where the superscript gives the order in the Bogoliubov expansion. Here $g = \langle b_l^\dagger b_l \rangle$ is the mean occupation number and we neglected depletion of the condensate. This is valid in the regime $J \gg U$ such that almost all atoms are in the zero momentum mode.

4.2 Light scattering

Light scattering by a one-dimensional optical lattice of ultracold atoms is studied in the setup sketched in Fig. 4.1. A laser plane wave at wave vector $\mathbf{k_L}$, frequency $\omega_L = c|\mathbf{k_L}|$, in a coherent state with amplitude α_L, drives the atoms. We evaluate the scattered light as a function of the angle of emission, determined by the wave vector $\mathbf{k}$ of the mode into which the photon is emitted, and of the frequency of the emitted photon.

The scattering process is evaluated assuming that the laser very weakly excites the atom, so that the atom-photon interaction is described at lowest order by Hamiltonian (4.1). More in detail, the condition $|\alpha_L| \ll 1$ means that the atomic sample is driven by at most one photon. A scattering process will then occur with probability $|\alpha_L|^2$ and will consist of the absorption of one incident photon in the mode of the laser, represented by the state $|1_L\rangle$, and the emission of a photon in one of the modes of the electromagnetic field at wave vector $\mathbf{k}$ and polarization $\epsilon_{\mathbf{k}} \perp \mathbf{k}$, represented by the state $|1_{\mathbf{k},\epsilon}\rangle$. The corresponding differential scattering cross section for the photon scattered at frequency ω in direction $\mathbf{n}$ in the solid angle Ω is proportional to the scattering rate (4.7) and takes the form [69]

$$\sigma(\Omega,\omega) = \frac{\mathcal{V}^2 \omega_L^2}{(2\pi)^2 \hbar^2 c^4} \sum_f \sum_{\epsilon_{\mathbf{k}} \perp \mathbf{n}} |\langle f, 1_{\mathbf{k},\epsilon}|H'_{\text{int}}|i, 1_L\rangle|^2 \delta^{(t/2)}(\omega_L + \omega_i - \omega - \omega_f), \quad (4.12)$$

where $\mathbf{k} = \mathbf{n}k$ and $|i\rangle, |f\rangle$ are the initial and final atomic states, eigenstates of Hamiltonian (4.2) at the eigenfrequencies ω_i and ω_f, respectively. Using Eq. (4.1) in Eq. (4.12) one can easily verify that the differential scattering cross section is proportional to the dynamic structure factor [21].

We evaluate the scattering cross section assuming that the atoms are initially in the ground state either of the Mott-insulator or of the superfluid phase, and that the atoms are scattered into a final state belonging to the lowest-lying atomic excitations. Using the form of operator H'_{int} in Eq. (4.3), Eq. (4.12) can be written as

where
$$\sigma(\Omega,\omega) = \sigma^{(0)}(\Omega,\omega) + \sigma^{(1)}(\Omega,\omega), \quad (4.13)$$

$$\sigma^{(0)} = \mathcal{A}(\Omega)\,|\langle i|\mathcal{T}(\mathbf{q})|i\rangle|^2\,\delta^{(t/2)}(\omega_L - \omega) \quad (4.14)$$

gives the elastic component of the scattered light, while

$$\sigma^{(1)} = \mathcal{A}(\Omega)\sum_f |\langle f, i_{\mathbf{k}}|\mathcal{T}(\mathbf{q})|i, 1_L\rangle|^2\,\delta^{(t/2)}(\omega_L - \omega - \delta\omega_f) \quad (4.15)$$

describes the scattering events in which one mechanical excitation at frequency $\delta\omega_f$ is absorbed from the photon by the atomic lattice (Stokes component) and corresponds to the one-phonon terms in neutron scattering [71]. The corresponding phonon emission processes, giving the anti-Stokes component, are here absent as initially the atoms are in the ground state. Moreover, higher order terms, corresponding to higher-order phonon terms in neutron scattering, are here neglected as we assume that at most one mechanical excitation is exchanged between lattice and photons.

The operator $\mathcal{T}(\mathbf{q})$ in the above equations is given in Eq. (4.4), while the coefficient $\mathcal{A}(\Omega)$ depends on the angle of emission and takes the form

$$\begin{aligned}\mathcal{A}(\Omega) &= \frac{\mathcal{V}^2 \omega_L^2}{(2\pi)^2 \epsilon_0^2 \hbar^2 c^4} \sum_{\epsilon_{\mathbf{k}} \perp \mathbf{n}} \frac{\hbar^2 |C_L^{e,g} C_{\mathbf{k}}^{e,g}|^2}{|\omega_L - \omega_0|^2} \\ &= \frac{\gamma}{c}\frac{\Omega_0^2}{\Delta^2}\left[\frac{3}{8\pi}\left(1 - \frac{|\mathbf{D}\cdot\mathbf{n}|^2}{|\mathbf{D}|^2}\right)\right],\end{aligned} \quad (4.16)$$

where γ is the linewidth of the dipole transition, $\Delta = \omega_L - \omega_0$ is the detuning of the laser from the atomic transition and $\Omega_0 = \sqrt{\omega_L/2\hbar\varepsilon_0}\,\mathbf{D}\cdot\boldsymbol{\epsilon}_L$.

4.2.1 Scattering cross section as a function of the atomic state

We now give an analytic expression for the scattering cross section in Eq. (4.12) for the initial and final states determined in Sec. 2.3.2 and 2.3.3.

Mott-insulator

For the Mott-insulator phase the initial state is $|i\rangle = |\psi_0^{(1)}\rangle$, given in Eq. (2.69). Using Eq. (4.3), we find

$$\sigma_{\text{MI}}^{(0)}(\Omega,\omega) = \mathcal{A}(\Omega)N^2\delta(\omega_L - \omega)\delta_{q_x,G}^{(M)}\left(|J_0(\mathbf{q})|^2 + 4\sqrt{g(g+1)}\frac{J}{U}\text{Re}\left\{J_0^*(\mathbf{q})J_1(\mathbf{q})\right\}\right), \quad (4.17)$$

4.2 Light scattering

where G are the vectors of the (one-dimensional) reciprocal lattice and

$$\delta_{q,G}^{(M)} \equiv \frac{1}{M^2} \frac{\sin^2(Md_0q/2)}{\sin^2(d_0q/2)} \qquad (4.18)$$

gives conservation of the Bloch momentum in a finite lattice with M sites, such that $\delta_{q,G}^{(M)} \to \delta_{q,G}$ (Kronecker delta) as $M \to \infty$. In Eq. (4.17) we omitted terms at third and higher order in J and $J_1(\mathbf{q})$. This approximation will be applied to the rest of this section, assuming that these higher-order terms can be neglected.

The presence of $\delta_{q_x,G}^{(M)}$ in Eq. (4.17) expresses the von-Laue condition for Bragg scattering. At zero order in the hopping term, Eq. (4.17) gives the response of a crystal of particles oscillating around their equilibrium position. In fact, using a Gaussian ansatz for the wave functions, one can estimate $|J_0(\mathbf{q})|^2 \simeq e^{-2W}$, with $W = [q_x^2\xi_x^2 + (q_y^2 + q_z^2)\xi_r^2]/8$, where ξ_x and ξ_r are the widths the atomic wave functions in the axial and radial direction, showing explicitly that this term is analogous to the Debye-Waller factor [71]. The term proportional to J is instead a novel feature with respect to traditional condensed-matter systems, that arises from light induced tunneling.

The Stokes component for the Mott-insulator is evaluated taking the final states $|f\rangle = |\psi_{[r,s]}^{(1)}\rangle$ given in Eq. (2.79), and reads

$$\sigma_{\text{MI}}^{(1)}(\Omega,\omega) = \mathcal{A}(\Omega) \sum_{r,s} \sin^2\left(\frac{\pi r}{M}\right) |\mathcal{B}_{r,s}|^2 \delta(\omega_L - \omega - \omega_{r,s}) \delta_{q(s),G}^{(M)}, \qquad (4.19)$$

with $\omega_{r,s} = (E_{r,s} - E_0)/\hbar$, and where we have introduced

$$q(s) = q_x - \frac{2\pi}{Md_0}s. \qquad (4.20)$$

The coefficient in Eq. (4.19) takes the form

$$\mathcal{B}_{r,s} = \sqrt{8g(g+1)} \begin{cases} J_1(\mathbf{q}) & \text{for } r+s \text{ odd}, \\ 2\frac{J}{U}J_0(\mathbf{q})\sin\left(\frac{\pi}{M}s\right) & \text{for } r+s \text{ even}, \end{cases} \qquad (4.21)$$

showing that the transition to the excited states with $r+s$ odd is due to photon recoil, and is hence a light-induced hopping process. Note that condition $q(s) = G$ shows that the quantum number s, and more specifically $2\pi s/L$, with $L = Md_0$ the length of the lattice, plays the role of the quasi-momentum of the states $|\psi_{r,s}^{(1)}\rangle$. We remark that Eq. (4.21), for $r+s$ even, agrees with the result evaluated in [21] (see Eq. (9) of that paper for comparison). The result we find for $r+s$ odd, on the contrary, is discarded in the treatment of [21], as there the authors neglected light induced hopping terms. In the Mott-insulator regime these terms are usually very small with respect to the other contributions. They give rise to a significant contribution when interfering with ordinary tunneling. This latter type of contributions is ruled out in the analytical model by the selection rule introduced by the assumption $g \gg 1$, but it is visible in the numerical results, as it will be shown in Sec. 4.2.

Superfluid

When evaluating the differential scattering cross section in the superfluid phase, we assume all atoms to be initially prepared in the Bose-Einstein condensate. In addition, for the analytical calculation we consider the limit $U \to 0$. In this limit we can neglect the quantum depletion of the condensate due to the interactions and take the initial state $|i\rangle = |0\rangle_{\mathrm{SF}}$ according to our notation. The zero-phonon term takes now the form

$$\begin{aligned}\sigma_{\mathrm{SF}}^{(0)}(\Omega,\omega) &= \mathcal{A}(\Omega)\delta(\omega_L - \omega)\delta_{q_x,G}^{(M)}N^2 \\ &\times \bigg(|J_0(\mathbf{q}) + 2J_1(\mathbf{q})|^2 \\ &\quad + 2\sum_{p\neq 0} \frac{|v_p|^2}{N}\mathrm{Re}\big\{(J_0(\mathbf{q}) + 2J_1(\mathbf{q}))^*(J_0(\mathbf{q}) + 2J_1(\mathbf{q})\cos(pd_0))\big\}\bigg), \end{aligned}$$

(4.22)

showing that the light-induced tunneling effects enter already at first order in this expression. As in the Mott-insulator case, the analogous of the Debye-Waller factor can be here identified in the term $|J_0(\mathbf{q})|^2$. In this case, though, tunneling effects become more important, modifying significantly the signal as we will show. The first-phonon term reads

$$\sigma_{\mathrm{SF}}^{(1)}(\Omega,\omega) = \mathcal{A}(\Omega)N\sum_{p\neq 0}\delta(\omega_L - \omega - \Omega_p)\frac{\epsilon_p}{\hbar\Omega_p}\left|\left(J_0(\mathbf{q}) + J_1(\mathbf{q})(1 + e^{-ipd_0})\right)\right|^2 \delta_{q_x-p,G}^{(M)},$$

(4.23)

and describes the creation of Bogoliubov excitations with quasi-momentum $\hbar p$ by photon scattering, such that the relation $p = q_x - G$ holds.

4.2.2 Numerical results

In this section we report the numerical results for the differential scattering cross section obtained when the atoms are in the Mott-insulator or in the superfluid state. The numerical results are obtained for a lattice of $M = 7$ sites and fixed particle number $N = M$. The coefficient entering the Bose-Hubbard Hamiltonian in Eq. (4.2) and the operator $\mathcal{T}(\mathbf{q})$ in Eq. (4.4) are calculated by using the Wannier functions relative to a given lattice depth V_0 of optical potential (2.26). Hamiltonian (4.2) is diagonalized exactly and the corresponding states are used for determining the differential scattering cross section in Eq. (4.12). The numerical results are also compared with the analytical predictions of the scattering cross sections reported in the previous section. Although the latter are valid for very large lattices and for large mean site occupation $g \gg 1$, we find reasonable agreement when comparing these predictions with those for a small lattice of 7 sites and single occupancy (see also [21]).

4.2 Light scattering

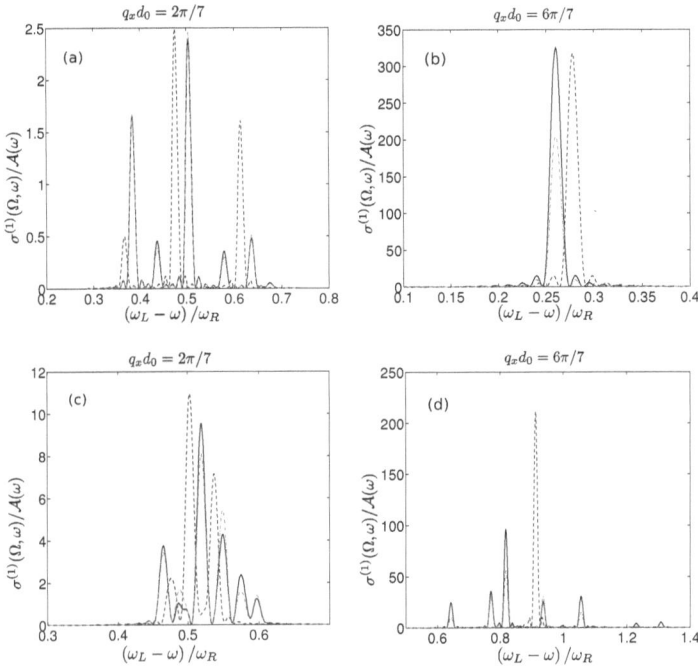

Figure 4.2: (color online) Stokes component of the differential scattering cross section (in units of $\mathcal{A}(\Omega)$) as a function of frequency (in units of the recoil frequency ω_R) for two different scattering angles, corresponding to $q_x d_0 = 2\pi/7$ (top row) and to $q_x d_0 = 6\pi/7$ (bottom row). The curves have been evaluated for a lattice of $M = 7$ site and $N = M = 7$ composed by ^{87}Rb atoms in the $|F = 2, m_F = 2\rangle$ hyperfine ground state. The black solid line corresponds to the numerical results, the blue dashed line to the analytical formulas (see text), the red dashed-dotted line to the model of [21], where the light-induced hopping is neglected. Plots (a) and (c) are evaluated for $V_0 = 8.1\hbar\omega_R$ ($U/J \approx 17$) which corresponds to the Mott-insulator state. Plots (b) and (d) are evaluated for $V_0 = 0.1\hbar\omega_R$ ($U/J \approx 1$) which corresponds to the superfluid state. Other parameters are $d_0 = 413$nm, $a_s = 105a_0$ with a_0 being the Bohr radius, and $\omega_r = 10\omega_R$ corresponding to the experimental parameters in [74] (For these parameters the size of the radial wavepacket is $\xi_r = 10a_s$). The frequency resolution is set to $\Delta\omega = 300$ Hz, corresponding to an integration time $T = 3$ msec.

Figure 4.2(a) and (c) display the one-phonon contribution to the differential scattering cross section, $\sigma^{(1)}(\Omega,\omega)$, as a function of the frequency ω and for different scattering angles when the atoms are in the Mott-insulator state. The numerical results are compared with the analytical model (dashed line) and with the model used in the numerical simulations in [21], in which light-induced hopping terms are not considered.

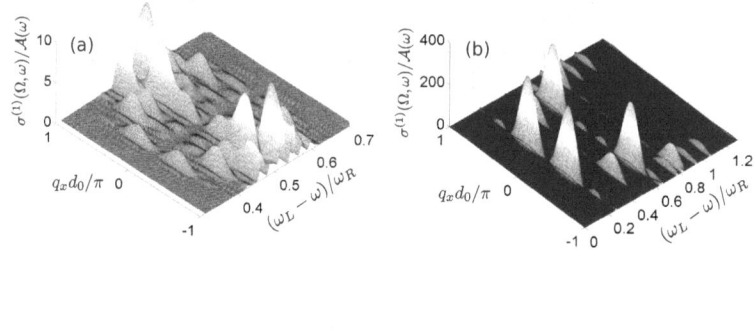

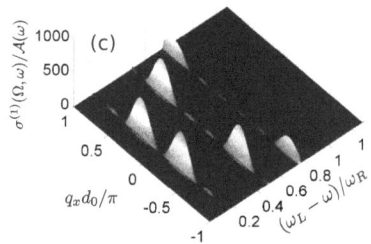

Figure 4.3: (color online) Stokes component of the differential scattering cross section (in units of $\mathcal{A}(\Omega)$) as a function of the frequency (in units of ω_R) and of the Bragg angle $\Theta = q_x d_0$ (in units of π). The plots have been evaluated numerically for (a) $V_0 = 8.1\hbar\omega_R$ and $U/J \approx 17$, (b) $V_0 = 0.1\hbar\omega_R$ and $U/J \approx 1$, (c) $V_0 = 0.1\hbar\omega_R$ and $U/J \approx 0.1$. The other parameters are as in Fig. 4.2.

We first discuss the numerical results which most closely approach the exact solution. The appearance of multiple peaks corresponds to the excitations of the atoms in the Mott-insulator due to the photon recoil. The number of peaks for the numerical

4.2 Light scattering

result is $M - 1$, which correspond in this case to 6. They can be individually resolved, as the system considered here is finite, and the width of each individual peak is limited by the detection time T (or the spectral resolution $1/T$). The linewidth of the excitations is essentially determined by the spontaneous decay rate of the excited state. In the present treatment we are assuming $\gamma'T \ll 1$, where γ' is the rate of incoherent scattering and is of the order $\gamma' \sim \gamma|C_L\alpha|^2/\Delta^2$. The analytical results are found using the model described in Sec. 2.3.3, which assumes a large on-site occupation. They are characterized by the same peak number, although only half of them is visible in the figure. In fact, the intensity of the peaks arising from the coupling of the ground state to the corresponding excitation via light-induced hopping (corresponding to the terms in Eq. (4.21) with $r + s$ odd) are very small compared to the other ones (corresponding to the terms with $r + s$ even) and are therefore not visible (note that, due to the assumption of large on-site occupation, interference between ordinary tunneling and light-induced hopping is suppressed). The central positions of the visible peaks present a systematic shift with respect to the ones found numerically. This systematic shift originates from the assumption $g \gg 1$, and has been observed in [21]. Nevertheless, the analytical solution still provides some insight into the numerical results. In Eq. (2.79), using Eq. (4.20) we find that the peaks are centered around the energy $E' = U$ with a spreading about this mean value of width $4J(2g+1)\cos\left(\frac{q_x d_0}{2}\right)$. Such spreading decreases as $q_x d_0$ approaches π, compare Fig. 4.2(a) and (c). In particular, for $q_x d_0 = \pi$, the width of the distribution of the Stokes excitations vanishes and the spectrum reduces to a single peak, corresponding to the on-site energy U.

The results for the superfluid regime are reported in Figs. 4.2(b) and (d). Here, the analytical solution predicts that in the limit $g \gg 1$ the total momentum of photon and lattice is conserved in a scattering event. Such property implies that the Bogoliubov mode matching the momentum-conservation condition, is excited, and therefore one expects a single peak in the spectrum. For $N = 7$ atoms and $g = 1$, the numerical results for $U/J \approx 1$ give a single peak at $q_x d_0 = 2\pi/7$, while at $q_x d_0 = 6\pi/7$ multiple peaks are found. In this case, instead of a collective density fluctuation with a well defined momentum p, one observes particle-hole types of excitations as in the Mott-insulator case. In Fig. 4.2(d) one observes a larger spread of the peaks as compared to the Mott-insulator case at the same Bragg angle. This is due to the larger value of the tunneling rate J. We remark that, choosing smaller values of the ratio U/J by ramping down the on-site interaction strength, as it is shown below, the spectrum reduces to a single peak at all Bragg angles and approaches the limit of the single-particle spectrum, as it is recovered in Eq. (2.66) by setting $U = 0$.

We now compare the numerical results, obtained taking systematically into account the light-induced hopping term, to the results found when this term is neglected, corresponding to the treatment in [21]. In the Mott-insulator case, comparison between the numerical results with and without light-induced hopping effects shows that in the first case one finds interference between ordinary tunneling and light-induced hopping. This

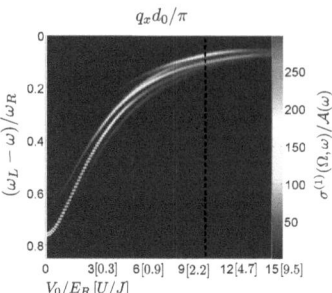

Figure 4.4: (color online) Contour plot of the Stokes component of the differential scattering cross section (in units of $\mathcal{A}(\Omega)$) as a function of the frequency (in units of ω_R) and of the lattice depth V_0 in units of ω_R for $q_x d_0 = 6\pi/7$ (the corresponding value of the ratio U/J is reported in the axis between squared bracket). The black dashed line marks the critical value at which the phase transition occurs in the thermodynamic limit. The other parameters are given in Fig. 4.2.

gives rise to an alternating enhancement and reduction of the peak heights at different frequencies, which is absent in the model discarding light-induced hopping effects. In general, the light-hopping term contributes in determining the height of some peaks, giving substantial modifications of the spectrum which can be revealed experimentally. The effect is larger in the superfluid regime, where tunneling is enhanced, as one can see in Fig. 4.2(b). Here, the central peak at $q_x d_0 = 2\pi/7$ is 50% higher than in absence of this contribution.

Figures 4.3(a)-(c) display the spectra of $\sigma^{(1)}$ as a function of the frequency and of the Bragg angle, in three different points of the phase diagram. We remark that the width and spacing of the Bragg peaks are determined by the finite size of the lattice. The plots in (a) and (b) are made in the same parameter regimes as in Fig. 4.2 (a),(c) and (b),(d), respectively, , namely $U/J \approx 17$ and $U/J \approx 1$. Figure 4.3(c), instead, corresponds to the value $U/J \sim 0.1$. Here, one observes almost a single peak at each Bragg angle, as expected in the weakly-interacting superfluid phase.

Figure 4.4 shows $\sigma^{(1)}$ as a function of the frequency and the depth of the potential, hence sweeping from the Mott-insulator to the superfluid regime at a given Bragg angle, corresponding to large momentum transfer ($q_x d_0 = 6\pi/7$). Here, one observes that the spectrum varies from multiple peaks, deep in the Mott-insulator regime, to a single peak in the weakly-interacting superfluid regime. The single peak appears around a value of U/J much smaller than the critical value $[U/J]_c$ for the Mott to superfluid transition (which, in the thermodynamic limit, is predicted for $[U/J]_c = 3.37$, see Ref. [115]). The presence of multiple peaks also in the superfluid phase close to the

4.2 Light scattering

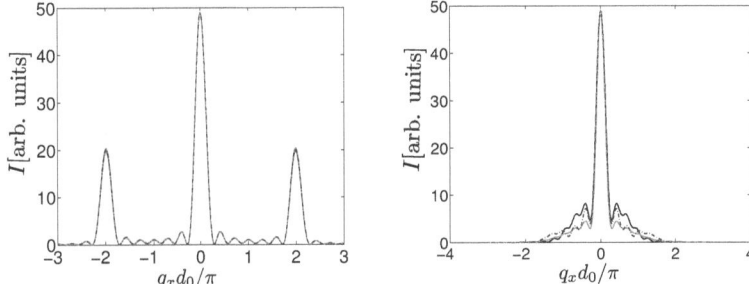

Figure 4.5: (color online) Intensity of the scattered light (in arbitrary units) as a function of the Bragg angle $0 - q_x d_0$ (in units of π). The parameters are the same as in Fig. 4.2 and (a) $V_0 = 8.1 E_R$ ($U/J \approx 17$), (b) $V_0 = 0.1 E_R$ ($U/J \approx 1$). The black solid line corresponds to the numerical result, the blue dashed-dotted line to the analytical solution, the red dashed line to the numerical result obtained discarding the light-induced hopping term as in [32, 21].

phase transition is reminiscent of a strongly-interacting superfluid phase. Such phase contains, beyond the gapless phononic modes, also gapped modes [116, 117, 79, 118, 119], which are predicted to be dominant at large quasi-momentum. We expect that also in the thermodynamic limit the transition to a single peak in the scattered-light spectrum will occur at lower values of U/J than the Mott-insulator to superfluid phase transition and will be also dependent on the momentum transfer. The identification of the Mott-insulator to superfluid phase transition should rather rely on the existence of a gapless spectrum. In spite of the very small size of the considered system, indications of a gapless spectrum are present in our results, as one can see comparing Fig. 4.3(a) with (b),(c).

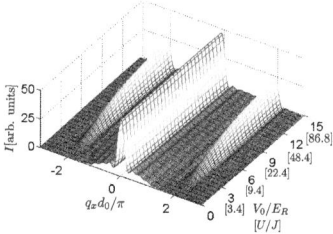

Figure 4.6: (color online) Intensity of the scattered light (in arbitrary units) as a function of the Bragg angle $\Theta = q_x d_0$ (in units of π) and of the lattice depth V_0 (in units of $\hbar \omega_R$) (the corresponding values of the ratio U/J are reported between squared brackets). The other parameters are reported in Fig. 4.2.

The intensity of the scattered light as a function of the Bragg angle is determined by the differential scattering cross section

$$\frac{d\sigma}{d\Omega} = \int d\omega \sigma(\Omega, \omega), \qquad (4.24)$$

and is reported in Figs. 4.5 for the atoms in (a) the Mott-insulator and in (b) the superfluid state. The solid line here corresponds to the numerical results, the dashed line to the analytical predictions and the dashed-dotted line to the model where light-induced hopping has been discarded, similar to the case considered in Ref. [32]. In this latter work, in fact, corrections due to the tunneling J were neglected when evaluating light scattering by the atoms in the Mott-insulator state, while the calculation of light scattering from the superfluid state was made discarding the finite value of the on-site interaction as well as the finite width of the Wannier functions. In Fig. 4.5(a) we observe that in the Mott-insulator regime the signal is dominated by the elastic component, and corresponds to a classical diffraction grating. In the superfluid regime, on the other hand, one finds that the amplitude of the Bragg peak is modified, and a background signal appears which is due to light scattering by the condensate fraction. This signal is the signature of the superfluid phase, and it arises from the coherent effects of tunneling and light-induced hopping. We also notice that in the superfluid phase only the first diffraction order is visible. This is due to the increased width of the atomic wave function, which yields a faster decaying Debye-Waller factor $J_0(\mathbf{q})$. We remark that higher diffraction orders would be visible if the superfluid regime was accessed by keeping the lattice depth constant, for instance by ramping down the on-site energy using a Feshbach resonance. The Bragg signal as a function of the lattice depth is reported in Fig. 4.6, showing the appearance of the background signal as the superfluid regime is approached.

4.3 Discussion

In this chapter we have discussed Bragg spectroscopy of ultracold bosonic atoms in an optical lattice, focussing on the signatures of the Mott-insulator and superfluid quantum state in the scattered photons. A full quantum theory for the atoms and photons dynamics and interactions has been developed, allowing us to identify the various contributions to the detected signals. We have characterized the Bragg scattering signal, for the parameters sweeping across the transition from the Mott-insulator to the superfluid quantum state. In particular, the contribution of light-induced hopping, arising from atomic recoil due to photon scattering, has been put into evidence. This term has been neglected in previous theoretical treatments [21, 32]. We have shown that its contribution can interfere with ordinary tunneling between sites thereby significantly affecting the spectroscopic signal. Its effect is visible in the behavior of the height of the peaks in the spectrum as a function of the emission angle, and it has been singled out by comparing the spectrum evaluated when this effect is discarded. This effect can be revealed experimentally in large systems, according to the analytical theory we develop by extending the one derived in [21, 77], and in small systems, as we observe by numerically evaluating the spectrum for a lattice of 7 atoms. It is interesting to consider whether such properties can be used as resources for photonic interfaces based on strongly-correlated atoms in optical lattices.

Measurement of the scattered light as a function of frequency at all angles in principle gives the information of the whole spectrum of the atoms and therefore completely determines the state of the atoms. In practice this is impossible, even in absence of all experimental imperferctions, due to finite resolution in frequency and angle.

As discussed in Sec. 4.2.2, the discrimination between Mott-insulator and superfluid phase in the present setup, should rely on the measurement of a finite gap in the excitation spectrum in the Mott-insulator state and its absence in the superfluid phase. This can be accomplished by measuring the inelastically scattered light close to the forward direction $\mathbf{q} = 0$. In the Mott-insulator there are no components of the inelastically scattered light at small **q** due to the gap in the excitation spectrum, whereas in the superfluid phase the components at small **q** are dominant as can be seen in Fig. (4.3) and Fig. (4.6). Measuring the inelasically scattered light at small **q** is experimentally very challenging since it would be difficult to distinguish between the elastic and inelastic component of the scattered light due to the small signal to noise ratio.

On the other hand, a clear signature of the phase transition from Mott-insulator to superfluid transition is the emergence of a nonzero value of the order paramterer $\langle b \rangle \neq 0$ in the superfluid phase. Measurement of the order parameter hence may reveal the many body state of the atoms. In the next chapter we will present a scheme that may allow one to measure the order parameter of an ultracold atomic gas and hence could be used for measuring the phase transition from Mott-insulator to superfluid phase.

Chapter 5

Atomic homodyning

Broken symmetry in statistical mechanics occurs when the Hamiltonian of a system is invariant under a symmetry operation but the ground state is not [120]. It is associated with the emergence of an order parameter, namely a physical quantity which is only nonzero in the state of broken symmetry. The concept of broken symmetry has a wide application in the description of physical phenomena, ranging from superfluidity of liquid helium [80, 121], superconductivity [80], Bose-Einstein condensation in weakly interacting atomic gases [122, 61], spontaneous magnetization of ferromagnets [120], till the Higgs boson in high energy physics [123, 124].

In cold atoms, second order phase transitions described by broken symmetry are for instance encountered in Bose-Einstein condensation of a weakly interacting atomic gas and in the Mott-insulator superfluid transition in optical lattices. In these cases the order parameter is the mean value of the atomic field operator $\langle\psi(\mathbf{r},t)\rangle$. The control parameter for the phase transition is the temperature in the case of Bose-Einstein condensation and the ratio between tunneling strength and onsite interaction energy in the Mott-insulator superfluid phase transition. A critical value of the control parameter separates the two possible phases.

In this chapter we discuss a method to measure $\langle\psi(\mathbf{r},t)\rangle$ via photodetection for a system of ultracold bosonic atoms which undergoes a spontaneous symmetry breaking. We first introduce the general setup where ultracold atoms are confined into two distinct regions in space and illuminated by two laser beams. We show that by monitoring the light intensity of one of the laser beams it is possible to acquire information about the mean value of the atomic field operators $\langle\psi(\mathbf{r},t)\rangle$ of the atomic systems. We apply our theory to the experimental setup of [47] where atoms were outcoupled from two independent Bose-Einstein condensates in order to measure their relative phase and reproduce the experimental results. We then demonstrate how an extension of the setup in [47] could be used in order to measure the temperature of a Bose-Einstein condensate and the superfluid fraction of ultracold atoms in an optical lattice. We end the chapter with a discussion of the results obtained.

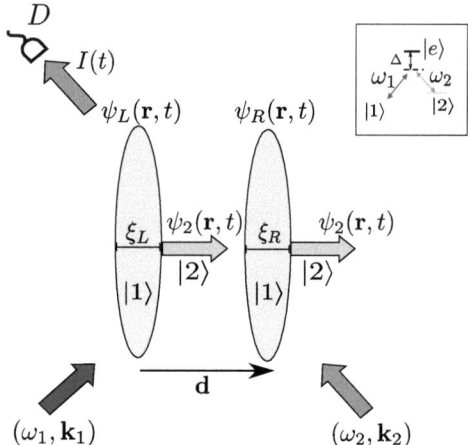

Figure 5.1: General setup for measuring the order parameter of a system $\langle\psi_R(\mathbf{r},t)\rangle$ when homodyning it with a reference system with known order parameter $\langle\psi_L(\mathbf{r},t)\rangle$. The relevant electronic levels of the atoms are in a lambda type configuration with two ground states $|1\rangle$ and $|2\rangle$. Atoms in state $|1\rangle$ are confined either in the left or right trap and described by the atomic field operators $\psi_L(\mathbf{r},t)$ and $\psi_R(\mathbf{r},t)$ respectively. The potential of the left (right) trap is assumed to be centered around $\mathbf{r}_L$ ($\mathbf{r}_R$) and $\mathbf{d} = \mathbf{r}_R - \mathbf{r}_L$. The typical radial size of the traps are denoted $\xi_{L,R}$. The traps are illuminated by two laser beams with frequency $\omega_{1,2}$ and wavevektor $\mathbf{k}_{1,2}$, where $a_{1,2}$ are the photonic annihilation operators for the two laser modes which drive a Raman transition between the atomic ground states $|1\rangle$ and $|2\rangle$. Atoms in state $|2\rangle$ do not feel any potential and are described by the field operator $\psi_2(\mathbf{r},t)$. Due to the photon recoil they will leave the traps as indicated by the green arrows.

5.1 The Model

We consider a gas of ultracold atoms trapped by an external potential whose functional form depends on the atoms electronic state. The atoms are assumed to be identical and with integer spin, hence obeying Bose-Einstein statistics. The electronic states of the atoms which are involved in the dynamics are a ground state, here denoted by the state $|1\rangle$, which is confined by the external potential $V_1(\mathbf{x})$, and a metastable state $|2\rangle$ in which the atomic center-of-mass motion does not experience forces. The two states are coupled via a dipolar transition to the common excited state $|e\rangle$ as shown in Fig. (1.1). An electronic transition between the two states permits hence one to switch the center-of-mass dynamics from atomic trapping to free evolution. We assume that the atoms are all initially prepared in the ground state $|1\rangle$ and that the potential confining the atoms in state $|1\rangle$ is splitted in two spatially separated regions centered at the points $\mathbf{r_L}$ and $\mathbf{r_R}$ with

$$\mathbf{d} = \mathbf{r}_R - \mathbf{r}_L \tag{5.1}$$

being the difference vector between the to trapping regions according to the scheme depicted in Fig. (5.1). We assume that there is no tunneling between the two regions, and moreover that the atoms in $|1\rangle$ confined to the left and right region can be considered as two independent systems to which we will refer in the following as left and right system respectively. We assume that the left and right systems exhibit no initial correlations with one another. The atoms in the left and right system are illuminated by a pair of laser beams, with wave vectors $\mathbf{k}_{1,2}$ and frequencies ω_{12}, which drive a two photon Raman transition between the states $|1\rangle$ and $|2\rangle$. The laser frequencies are in resonance for the Raman coupling between the states $|1\rangle$ and $|2\rangle$ and far detuned from the excited state $|e\rangle$, so that radiative decay can be neglected. In this case the excited state can be adiabatically eliminated using the procedure described in Sec. 2.1 and the dynamics of the system is governed by the Hamiltonian

$$H = H_0 + H_{emf} + V, \tag{5.2}$$

where the term

$$H_{emf} = \hbar\omega_1 a_1^\dagger a_1 + \hbar\omega_2 a_2^\dagger a_2 \tag{5.3}$$

$$\tag{5.4}$$

describes the energy of the free electromagnetic field taking only into account the two laser modes. The operator $a_{1,2}$ ($a_{1,2}^\dagger$) is the annihilation (creation) operator for a photon in the laser mode $\omega_{1,2}$. The term

$$H_0 = \sum_{j=L,R,2} (H_j + \varepsilon_j N_j) \tag{5.5}$$

gives the unperturbed atomic part. In Eq. (5.5) $j = L$ ($j = R$) refers to atoms in the left (right) system, ε_j denotes the internal energy of the atoms with $\varepsilon_L = \varepsilon_R = \varepsilon_1$

and $N_j = \int d\mathbf{r} \psi_j^\dagger(\mathbf{r})\psi_j(\mathbf{r})$ is the number operator for atoms in left (right) trap and the outcoupled atoms. The operator $\psi_j(\mathbf{r})$ ($\psi_j^\dagger(\mathbf{r})$) annihilates (creates) an atom in the left ($j = L$), right ($j = R$) system or an outcoupled atom ($j = 2$) at position $\mathbf{r}$. The Hamiltonian describing the atomic dynamics H_j is given by

$$H_{j=L,R} = \int d\mathbf{r} \psi_j^\dagger(\mathbf{r}) \left(\frac{-\hbar^2 \nabla^2}{2m} + V_j(\mathbf{r}) \right) \psi_j(\mathbf{r}) + \frac{g}{2} \int d\mathbf{r} \psi_j^\dagger(\mathbf{r})\psi_j^\dagger(\mathbf{r})\psi_j(\mathbf{r})\psi_j(\mathbf{r}) \quad (5.6a)$$

$$H_2 = \int d\mathbf{r} \psi_2^\dagger(\mathbf{r}) \left(\frac{-\hbar^2 \nabla^2}{2m} \right) \psi_j(\mathbf{r}). \quad (5.6b)$$

where g determines the strength of s-wave scattering between the atoms in state $|1\rangle$ and is defined in Eq. (1.23). Atomic collisions between outcoupled atoms in state $|2\rangle$ can be neglected due to the low density of outcoupled atoms. We will also neglect collisions between outcoupled atoms and trapped atoms in state $|1\rangle$[1].

The effective light-matter interaction after adiabatic elimination of state $|e\rangle$ is derived from Eq. (2.2) and reads

$$V = \hbar \int d\mathbf{r} \left[\gamma_0 e^{i\mathbf{q}\cdot\mathbf{r}} \psi_2^\dagger(\mathbf{r})(\psi_L(\mathbf{r}) + \psi_R(\mathbf{r})) a_2^\dagger a_1 + h.c. \right]. \quad (5.7)$$

Here $\mathbf{q} = \mathbf{k}_1 - \mathbf{k}_2$ is the difference in wave vectors of the two laser beams and

$$\gamma_0 = \frac{|C_1^{e,1} C_2^{e,2}|}{\Delta} \quad (5.8)$$

is the coupling constant, where Δ is the detuning from the excited state $|e\rangle$ as indicated in Fig. (5.1) and C_j^j is defined in Eq. (1.53). In writing Eq. (5.7) we have neglected scattering into modes other than the laser modes. We note that the system we consider is an extension to the one in [125, 126], where the authors studied the outcoupling of atoms from a single Bose-Einstein condensate with classical Raman lasers. Here we treat the light quantum mechanically and show that the scattered light can give access to the properties of the scattering system.

5.2 Scattered light intensity

The quantity we will analyse is the light intensity $I(t)$ at the detector D as sketched in Fig. (5.1). It can be written as the sum of two contributions

$$I(t) = I_0 + I_V(t). \quad (5.9)$$

where I_0 is the background light intensity, that is the intensity of the laser at frequency ω_2 and wavevector $\mathbf{k}_2$. The quantity $I_V(t)$ is the contribution of the photons scattered

---
[1] Interaction between the trapped and outcoupled atoms can be made small by means of a Feschbach resonance, where by means of an external magnetic field one can tune the s-wave scattering length [14].

5.2 Scattered light intensity

by the atomic medium from the laser mode at frequency ω_1 into the laser mode at frequency ω_2 via coherent Raman processes as indicated in Fig. (5.1) and is given by

$$I_V(t) \propto \frac{N_V(t)}{\Delta t}.\tag{5.10}$$

Here Δt is the integration time of the detector and $N_V(t)$ is the number of photons scattered by the atoms from the laser mode at frequency ω_1 into the laser mode at frequency ω_2 in the time interval $[t - \Delta t/2, t + \Delta t/2]$. We thereby neglect retardation effects due to the time a photon needs to travel the distance between the scattering atom and the detector. $N_V(t)$ can be written as

$$N_V(t) = \int_{t-\Delta t/2}^{t+\Delta t/2} dt' F(t'),\tag{5.11}$$

where the photon flux

$$F(t) = \frac{d}{dt}\langle a_2^\dagger a_2 \rangle \tag{5.12}$$

is the rate of change of the photon number in the laser mode at frequency ω_2. Using Eq. (5.11) in Eq. (5.10) we find

$$I_V(t) \propto \frac{N_V(t)}{\Delta t} = \frac{1}{\Delta t}\int_{t-\Delta t/2}^{t+\Delta t/2} F(t')dt' \tag{5.13a}$$
$$\approx F(t),\tag{5.13b}$$

where in Eq. (5.13b) we have assumed that the integration time of the detector is much smaller than the typical time scale for variations of the photon flux $F(t)$. We will return to this point in Sec. 5.3 where we check its validity for specific setups. According to Eq. (5.9) and Eq. (5.13) the photon flux is directly proportional to the intensity measured at the photodetector after subtracting the background signal,

$$F(t) \propto I(t) - I_0.\tag{5.14}$$

We now derive the explicit form of the photon flux $F(t)$ and show that it provides a measure of the mean value of the atomic field operator. We calculate $F(t)$ up to second order in V according to standard perturbation theory as shown in Chapter 1. The detailed steps are reported in Appendix A. From the form of the equation for $F(t)$ we find that the photon flux can be decomposed into the sum of two contributions according to the decomposition

$$F(t) = F_B(t) + F_I(t),\tag{5.15}$$

whereby

$$F_B(t) = \Gamma \operatorname{Re} \int_0^t dt' \int d\mathbf{r}\, d\mathbf{r}'\, e^{i(\mathbf{q}\cdot(\mathbf{r}-\mathbf{r}')-\omega_{12}(t-t'))} \left\langle \psi_2(\mathbf{r}',t')_{H_0}\psi_2^\dagger(\mathbf{r},t)_{H_0}\right\rangle \quad (5.16a)$$
$$\times \left(\left\langle \psi_L^\dagger(\mathbf{r}',t')_{H_0}\psi_L(\mathbf{r},t)_{H_0}\right\rangle + \left\langle \psi_R^\dagger(\mathbf{r}',t')_{H_0}\psi_R(\mathbf{r},t)_{H_0}\right\rangle \right),$$

$$F_I(t) = \Gamma \operatorname{Re} \int_0^t dt' \int d\mathbf{r}\, d\mathbf{r}'\, e^{i(\mathbf{q}\cdot(\mathbf{r}-\mathbf{r}')-\omega_{12}(t-t'))} \left\langle \psi_2(\mathbf{r}',t')_{H_0}\psi_2^\dagger(\mathbf{r},t)_{H_0}\right\rangle \quad (5.16b)$$
$$\times \left(\left\langle \psi_R^\dagger(\mathbf{r}',t')_{H_0}\psi_L(\mathbf{r},t)_{H_0}\right\rangle + \left\langle \psi_L^\dagger(\mathbf{r}',t')_{H_0}\psi_R(\mathbf{r},t)_{H_0}\right\rangle \right),$$

where we introduced the coupling constant

$$\Gamma = 2|\gamma_0|^2 n_1(n_2+1), \quad (5.17)$$

and the difference frequency

$$\omega_{12} = \omega_1 - \omega_2. \quad (5.18)$$

In Eq. (5.16) all operators are in the Heisenberg picture with respect to H_0. Moreover we made use of the fact that initially there are no atoms in state $|2\rangle$. Hence the expectation value for the field operators of the atoms in state $|2\rangle$ taken over the vacuum yields

$$\langle \psi_2^\dagger(\mathbf{r}',t')_{H_0}\psi_2^\dagger(\mathbf{r},t)_{H_0}\rangle = \langle \psi_2^\dagger(\mathbf{r}',t')_{H_0}\psi_2(\mathbf{r},t)_{H_0}\rangle = 0.$$

We note that term $F_B(t)$, Eq. (5.16a), is the sum of the contribution of each atomic system to the Raman scattering rate. Term $F_I(t)$, Eq. (5.16b), can be interpreted as describing processes where an atom in transferred, say, from the left to the right trap through the coupling via the freely propagating state $|2\rangle$. We will discuss in detail the implications of this term in Sec. (5.2.2). In the following we will refer to $F_B(t)$ as the background contribution and to $F_I(t)$ as the interference contribution to the photon flux.

In order to evaluate the photon flux in terms of atomic correlation functions it is convenient to perform the interaction picture with respect to the Hamiltonian for the grand-canonical ensemble

$$K_0 = H_0 - \sum_{j=L,R,2}(\epsilon_j + \mu_j)N_j, \quad (5.19)$$

where the chemical potential for the outcoupled atoms is zero $\mu_2 = 0$ and the operators in this picture read

$$A_{K_0} = U_{K_0}^\dagger(t) A U_{K_0}(t), \quad (5.20a)$$
$$U_{K_0}(t) = e^{-\frac{i}{\hbar}K_0 t}. \quad (5.20b)$$

5.2 Scattered light intensity

We use $e^{\lambda N_j}\psi_k(\mathbf{r})e^{-\lambda N_j} = e^{-\lambda \delta_{j,k}}\psi_j(\mathbf{r})$ to connect the original field operators in the usual Heisenberg picture and the field operators in the grand-canonical ensemble

$$\begin{aligned}\psi_k(\mathbf{r},t)_{H_0} &= e^{\frac{i}{\hbar}H_0 t}\psi_k(\mathbf{r})e^{-\frac{i}{\hbar}H_0 t}\\ &= e^{\frac{i}{\hbar}K_0 t}e^{\frac{i}{\hbar}\sum_j(\mu_j+\varepsilon_j)N_j t}\psi_k(\mathbf{r})e^{-\frac{i}{\hbar}\sum_j(\mu_j+\varepsilon_j)N_j t}e^{-\frac{i}{\hbar}K_0 t}\\ &= e^{-\frac{i}{\hbar}(\mu_k+\varepsilon_k)t}\psi_k(\mathbf{r},t)_{K_0},\end{aligned} \quad (5.21)$$

where the subscripts indicate in which Heisenberg picture the operators have to be taken. Using Eq. (5.21) the contributions to the photon flux Eq. (5.16) are rewritten as

$$F_B(t) = \Gamma\operatorname{Re}\int_0^t dt' \int d\mathbf{r}\, d\mathbf{r}' e^{i(\mathbf{q}\cdot(\mathbf{r}-\mathbf{r}')-\Omega(t-t'))} \left\langle \psi_2(\mathbf{r}',t')_{K_0} \psi_2^\dagger(\mathbf{r},t)_{K_0} \right\rangle \quad (5.22a)$$
$$\times \left(\left\langle \psi_L^\dagger(\mathbf{r}',t')_{K_0}\psi_L(\mathbf{r},t)_{K_0}\right\rangle + e^{i\delta\mu(t-t')}\left\langle \psi_R^\dagger(\mathbf{r}',t')_{K_0}\psi_R(\mathbf{r},t)_{K_0}\right\rangle \right),$$

$$F_I(t) = \Gamma\operatorname{Re}\int_0^t dt' \int d\mathbf{r}\, d\mathbf{r}' e^{i(\mathbf{q}\cdot(\mathbf{r}-\mathbf{r}')-\Omega(t-t'))} \left\langle \psi_2(\mathbf{r}',t')_{K_0} \psi_2^\dagger(\mathbf{r},t)_{K_0} \right\rangle \quad (5.22b)$$
$$\times \left(e^{-i\delta\mu t'}\left\langle \psi_R^\dagger(\mathbf{r}',t')_{K_0}\psi_L(\mathbf{r},t)_{K_0}\right\rangle + e^{i\delta\mu t}\left\langle \psi_L^\dagger(\mathbf{r}',t')_{K_0}\psi_R(\mathbf{r},t)_{K_0}\right\rangle \right),$$

where

$$\hbar\Omega = \hbar\omega_{12} + \mu_L - \mu_2 + \varepsilon_1 - \varepsilon_2, \quad (5.23a)$$
$$\delta\mu = \frac{\mu_L - \mu_R}{\hbar}. \quad (5.23b)$$

In the following we will omit the subscripts K_0 and implicit take all atomic operators in the interaction picture with respect to K_0

Since there are no initial correlations between the left and right system we can split the expectation value

$$\left\langle \psi_R^\dagger(\mathbf{r}',t')\psi_L(\mathbf{r},t)\right\rangle = \left\langle \psi_R^\dagger(\mathbf{r}',t')\right\rangle\left\langle \psi_L(\mathbf{r},t)\right\rangle.$$

Thus the interference flux $F_I(t)$ in Eq. (5.22b) is proportional to the product of the mean values of the atomic field operators and is nonvanishing only if there is a superfluid component in both systems.

While the form of the expectation values $\langle \psi_{L,R}^\dagger(\mathbf{r}',t')\psi_{L,R}(\mathbf{r},t)\rangle$ for the trapped atoms depend on the specific initial state, the expectation value $\langle \psi_2^\dagger(\mathbf{r}',t')\psi_2(\mathbf{r},t)\rangle$ is the propagator of a free particle with mass m and can be evaluated by expanding the atomic field operators in momentum modes

$$\psi_2(\mathbf{r},t) = \frac{1}{\sqrt{V}}\sum_{\mathbf{p}} e^{\frac{i}{\hbar}[\mathbf{p}\cdot\mathbf{r}-\frac{\mathbf{p}^2}{2m}t]}a_{\mathbf{p}}. \quad (5.24)$$

Here $a_\mathbf{p}$ annihilates an atom with momentum $\mathbf{p}$ and the propagator for the outcoupled atoms can be written as

$$\begin{aligned}\left\langle \psi_2(\mathbf{r}',t')\psi_2^\dagger(\mathbf{r},t)\right\rangle &= \frac{1}{V}\sum_\mathbf{p} e^{\frac{i}{\hbar}[\mathbf{p}\cdot(\mathbf{r}'-\mathbf{r})-\frac{\mathbf{p}^2}{2m}(t'-t)]}\\ &= \int\frac{d\mathbf{k}}{(2\pi)^3}e^{i[\mathbf{k}\cdot(\mathbf{r}'-\mathbf{r})-\omega_\mathbf{k}(t'-t)]}\,,\end{aligned} \quad (5.25\text{a})$$

where we replaced the sum over all momenta by an integral in the standard way assuming large volumes V and we introduced $\omega_\mathbf{k} = \frac{\hbar \mathbf{k}^2}{2m}$. The integral in Eq. (5.25a) gives

$$\left\langle \psi_2(\mathbf{r}',t')\psi_2^\dagger(\mathbf{r},t)\right\rangle = \left(\frac{m}{2\pi i\hbar(t'-t)}\right)^{\frac{3}{2}}\exp\left[\frac{i}{\hbar}\frac{m(\mathbf{r}'-\mathbf{r})^2}{2(t'-t)}\right]\,. \quad (5.25\text{b})$$

The propagator for the outcoupled atoms given in the form Eq. (5.25b) is usually encountered in the context of Feynman path integrals [127].

Using Eq. (5.25a) in Eq. (5.22) the two contributions to the photon flux read

$$\begin{aligned}F_B(t) &= \Gamma\text{Re}\int_0^t dt'\int\frac{d\mathbf{k}}{(2\pi)^3}\int d\mathbf{r}\,d\mathbf{r}'e^{i(\mathbf{q}-\mathbf{k})\cdot(\mathbf{r}-\mathbf{r}')}e^{-i(\Omega-\omega_\mathbf{k})(t-t')}\\ &\quad \times \left(\left\langle \psi_L^\dagger(\mathbf{r}',t')\psi_L(\mathbf{r},t)\right\rangle + e^{i\delta\mu(t-t')}\left\langle \psi_R^\dagger(\mathbf{r}',t')\psi_R(\mathbf{r},t)\right\rangle\right)\,,\end{aligned} \quad (5.26\text{a})$$

$$\begin{aligned}F_I(t) &= \Gamma\text{Re}\int_0^t dt'\int\frac{d\mathbf{k}}{(2\pi)^3}\int d\mathbf{r}\,d\mathbf{r}'e^{i(\mathbf{q}-\mathbf{k})\cdot(\mathbf{r}-\mathbf{r}')}e^{-i(\Omega-\omega_\mathbf{k})(t-t')}\\ &\quad \times \left(e^{-i\delta\mu t'}\left\langle \psi_R^\dagger(\mathbf{r}',t')\psi_L(\mathbf{r},t)\right\rangle + e^{i\delta\mu t}\left\langle \psi_L^\dagger(\mathbf{r}',t')\psi_R(\mathbf{r},t)\right\rangle\right)\,. \end{aligned} \quad (5.26\text{b})$$

5.2.1 Background contribution

We now evaluate the background contribution to the photon flux and first assume that the atomic systems are harmonically trapped Bose-Einstein condensates. We split the atomic field operator into a part that describes the Bose-Einstein condensate by means of a macroscopic wave function and a part that takes the noncondensed atoms into account [61]

$$\psi_j(\mathbf{r},t) = \Phi_j(\mathbf{r}) + \delta\psi_j(\mathbf{r},t)\,. \quad (5.27)$$

Here $\Phi_j(\mathbf{r}) = \langle\psi_j(\mathbf{r},t)\rangle$ denotes the macroscopic wave function and $\delta\psi_j(\mathbf{r},t)$ describes the noncondensed atoms with $\langle\delta\psi_j(\mathbf{r},t)\rangle = 0$. We note that the macroscopic wave function $\Phi_j(\mathbf{r})$ is independent on time in the interaction picture with respect to the Hamiltonian of the grand canonical ensemble. For later convenience we write it as

$$\Phi_j(\mathbf{r}) = f_j(\mathbf{r}-\mathbf{r}_j)e^{i\varphi_j}\,, \quad (5.28)$$

with both $f_j(\mathbf{r})$ and φ_j real. The vector $\mathbf{r}_j$ points to the trap minima of system $j = (L,R)$ and $f_j(\mathbf{r}-\mathbf{r}_j)^2$ is the density of condensed atoms for system j. The phase φ_j is

5.2 Scattered light intensity

the macroscopic phase of the condensate in system j and we assume it to be constant in space, hence discarding the possibility of superfluid currents in the left and right Bose-Einstein condensate [61]. Using Eq. (5.27) and Eq. (5.28) in Eq. (5.26a) leads to identify four contributions to the background signal

$$F_B(t) = F_L(t) + F_{L,Q}(t) + F_R(t) + F_{R,Q}(t), \tag{5.29}$$

where

$$F_L(t) = \pi\Gamma \int \frac{d\mathbf{k}}{(2\pi)^3} |f_L(\mathbf{q}-\mathbf{k})|^2 \delta^t(\Omega - \omega_\mathbf{k}), \tag{5.30a}$$

$$F_R(t) = \pi\Gamma \int \frac{d\mathbf{k}}{(2\pi)^3} |f_R(\mathbf{q}-\mathbf{k})|^2 \delta^t(\Omega - \omega_\mathbf{k} - \delta\mu), \tag{5.30b}$$

are the contributions from the macroscopic wave functions of the left and right system respectively with $f_j(\mathbf{k}) = \int d\mathbf{r} e^{-i\mathbf{k}\cdot\mathbf{r}} f_j(\mathbf{r})$, such that $|f_j(\mathbf{k})|^2$ is the momentum distribution of the macroscopic wavefunction and $\delta^t(\omega)$ the diffraction function enforcing energy conservation for long interaction times t, see Eq. (4.7). Let us now discuss the physical meaning of Eqs. (5.30). The integrals run over all atomic momenta and are weighted by the momentum distribution at $\mathbf{k}' = \mathbf{q} - \mathbf{k}$ and the diffraction function, hence accounting for the photon recoil due to the outcoupling process and energy conservation. They can thus be seen as the total number of condensed atoms available to the outcoupling process, due to the restrictions given by energy and momentum conservation in photon scattering.

The remaining two terms $F_{j,Q}(t)$ in Eq. (5.29) are due to scattering by noncondensed atoms and read

$$\begin{aligned}F_{L,Q}(t) = \Gamma\mathrm{Re}\int \frac{d\omega}{2\pi} \int_0^t d\tau \int \frac{d\mathbf{k}}{(2\pi)^3} e^{-i(\omega+\Omega-\omega_\mathbf{k})\tau} \\ \times \int d\mathbf{r}\, d\mathbf{r}' e^{i(\mathbf{q}-\mathbf{k})\cdot(\mathbf{r}-\mathbf{r}')} \left\langle \delta\psi_L^\dagger(\mathbf{r}',-\omega)\delta\psi_L(\mathbf{r})\right\rangle,\end{aligned} \tag{5.31a}$$

$$\begin{aligned}F_{R,Q}(t) = \Gamma\mathrm{Re}\int \frac{d\omega}{2\pi} \int_0^t d\tau \int \frac{d\mathbf{k}}{(2\pi)^3} e^{-i(\omega+\Omega-\omega_\mathbf{k}-\delta\mu)\tau} \\ \times \int d\mathbf{r}\, d\mathbf{r}' e^{i(\mathbf{q}-\mathbf{k})\cdot(\mathbf{r}-\mathbf{r}')} \left\langle \delta\psi_R^\dagger(\mathbf{r}',-\omega)\delta\psi_R(\mathbf{r})\right\rangle.\end{aligned} \tag{5.31b}$$

We have used the equality

$$\left\langle \delta\psi_j^\dagger(\mathbf{r}',t)\delta\psi_j(\mathbf{r})\right\rangle = \int \frac{d\omega}{2\pi} e^{-i\omega t} \left\langle \delta\psi_j^\dagger(\mathbf{r}',\omega)\delta\psi_j(\mathbf{r})\right\rangle \tag{5.32}$$

in Eq. (5.31), thereby expressing these correlation functions in terms of their Fourier transforms. The expectation values in Eq. (5.31) can be rewritten as [80]

$$\left\langle \delta\psi_j^\dagger(\mathbf{r}',-\omega)\delta\psi_j(\mathbf{r})\right\rangle = N_0(\omega) A_{\delta\psi_j \delta\psi_j^\dagger}(\mathbf{r},\mathbf{r}',\omega), \tag{5.33}$$

where
$$N_0(\omega) = \frac{1}{e^{\beta\hbar\omega} - 1} \qquad (5.34)$$

is the Bose function and $A_{\tilde{\psi}^\dagger\tilde{\psi}}(\mathbf{r}', \mathbf{r}, \omega)$ is the spectral density of the excitations, which reads

$$A_{\delta\psi\delta\psi^\dagger}(\mathbf{r}, \mathbf{r}', \omega) = \int_{-\infty}^{\infty} dt\, e^{i\omega t} \left\langle \left[\delta\psi(\mathbf{r},t), \delta\psi^\dagger(\mathbf{r}')\right]\right\rangle. \qquad (5.35)$$

Using Eq. (5.33) in Eq. (5.31) we find

$$\begin{aligned}
F_{L,Q}(t) &= \Gamma\mathrm{Re}\int\frac{d\omega}{2\pi}\int_0^t d\tau \int \frac{d\mathbf{k}}{(2\pi)^3} e^{-i(\omega+\Omega-\omega_{\mathbf{k}})\tau}\int d\mathbf{r}\,d\mathbf{r}'\\
&\quad \times e^{i(\mathbf{q}-\mathbf{k})\cdot(\mathbf{r}-\mathbf{r}')} N_0(\omega) A_{\delta\psi_L \delta\psi_L^\dagger}(\mathbf{r},\mathbf{r}',\omega), & (5.36a)\\
F_{R,Q}(t) &= \Gamma\mathrm{Re}\int\frac{d\omega}{2\pi}\int_0^t d\tau \int \frac{d\mathbf{k}}{(2\pi)^3} e^{-i(\omega+\Omega-\omega_{\mathbf{k}}-\delta\mu)\tau}\int d\mathbf{r}\,d\mathbf{r}'\\
&\quad \times e^{i(\mathbf{q}-\mathbf{k})\cdot(\mathbf{r}-\mathbf{r}')} N_0(\omega) A_{\delta\psi_R \delta\psi_R^\dagger}(\mathbf{r},\mathbf{r}',\omega). & (5.36b)
\end{aligned}$$

In order to get an intuitive understanding of Eqs. (5.36) we assume that the spatial variation of the systems can be neglected such that we can assume the systems to be homogeneous[2]. This leads to $A_{\delta\psi_j \delta\psi_j^\dagger}(\mathbf{r},\mathbf{r}',\omega) = A_{\delta\psi_j \delta\psi_j^\dagger}(|\mathbf{r}-\mathbf{r}'|,\omega)$, and Eqs. (5.36) can be written as

$$\begin{aligned}
F_{L,Q}^{\mathrm{hom}}(t) &= \pi\Gamma V \int\frac{d\omega}{2\pi}\int \frac{d\mathbf{k}}{(2\pi)^3} \delta^t(\omega+\Omega-\omega_{\mathbf{k}}) N_0(\omega) A_{\delta\psi_L \delta\psi_L^\dagger}(\mathbf{k}-\mathbf{q},\omega), & (5.37a)\\
F_{R,Q}^{\mathrm{hom}}(t) &= \pi\Gamma V \int\frac{d\omega}{2\pi}\int \frac{d\mathbf{k}}{(2\pi)^3} \delta^t(\omega+\Omega-\omega_{\mathbf{k}}-\delta\mu) N_0(\omega) A_{\delta\psi_R \delta\psi_R^\dagger}(\mathbf{k}-\mathbf{q},\omega). & \\
& & (5.37b)
\end{aligned}$$

where we use the superscript "hom" in order to make our assumption explicit. In this form we can provide a physical interpretation of Eq. (5.37). For a homogeneous system the spectral weight function $A_{\delta\psi_R \delta\psi_R^\dagger}(\mathbf{k},\omega)$ gives the strength of the collective excitations at wavevector $\mathbf{k}$ and frequency ω, while the Bose function $N_0(\omega)$ gives their thermal weight. The total contribution for the outcoupling process due to the quantum fluctuation is thus the total number of the collective excitations obeying energy and momentum conservation.

The results in Eq. (5.36) and Eq. (5.30) for the background current have been derived by Luxat *et al.* [125] who calculated the stream of outcoupled atoms from a

---

[2] Assuming e.g. strong confinement for the gases in the x-direction such that the motional degrees of the atoms are restricted to the y and z-direction one could realize two parallel homogeneous atomic sheets with no overlap

5.2 Scattered light intensity

single Bose-Einstein condensate at finite temperature modeling the outcoupling process by two classical Raman lasers. Indeed from Eq. (5.7) one can easily show that

$$\frac{d}{dt}\langle a_2^\dagger a_2 \rangle = \frac{d}{dt} N_2, \qquad (5.38)$$

where N_2 is the number of outcoupled atoms. Thus the quantity $N_2 - \langle a_2^\dagger a_2 \rangle$ is conserved during the outcoupling process, showing that for each outcoupled atom there is an additional photon in the laser mode at frequency ω_2. Measuring the outcoupled atom flux or the Raman scattering rate of the photons provide the same information about the trapped atomic system.

5.2.2 Interference contribution

In order to calculate the term $F_I(t)$ we split it into the sum

$$F_I(t) = F_{L \to R}(t) + F_{R \to L}(t), \qquad (5.39)$$

with

$$\begin{aligned}
F_{L \to R}(t) &= \Gamma \operatorname{Re} e^{i\delta\mu t} \int_0^t dt' \int \frac{d\mathbf{k}}{(2\pi)^3} e^{-i(\Omega - \omega_\mathbf{k})(t-t')} \\
&\quad \times \int d\mathbf{r}\, d\mathbf{r}'\, e^{i(\mathbf{q}-\mathbf{k})\cdot(\mathbf{r}-\mathbf{r}')} \Phi_L(\mathbf{r}')^* \Phi_R(\mathbf{r}), \qquad (5.40) \\
F_{R \to L}(t) &= \Gamma \operatorname{Re} e^{-i\delta\mu t} \int_0^t dt' \int \frac{d\mathbf{k}}{(2\pi)^3} e^{-i(\Omega - \omega_\mathbf{k} - \delta\mu)(t-t')} \\
&\quad \times \int d\mathbf{r}\, d\mathbf{r}'\, e^{i(\mathbf{q}-\mathbf{k})\cdot(\mathbf{r}-\mathbf{r}')} \Phi_R(\mathbf{r}')^* \Phi_L(\mathbf{r}). \qquad (5.41)
\end{aligned}$$

We note that the noncondensed atoms do not contribute to the interference term since $\langle \delta\psi(\mathbf{r},t) \rangle = 0$. Using Eq. (5.28) they can be written as

$$\begin{aligned}
F_{L \to R}(t) &= \Gamma \operatorname{Re} \int_0^t dt' \int \frac{d\mathbf{k}}{(2\pi)^3} e^{-i(\Omega - \omega_{\mathbf{k}+\mathbf{q}})(t-t')} \\
&\quad \times e^{i(\delta\mu t - \varphi_{LR} - \mathbf{k}\cdot\mathbf{d})} f_L(\mathbf{k})^* f_R(\mathbf{k}), \qquad (5.42a) \\
F_{R \to L}(t) &= \Gamma \operatorname{Re} \int_0^t dt' \int \frac{d\mathbf{k}}{(2\pi)^3} e^{-i(\Omega - \omega_{\mathbf{k}+\mathbf{q}} - \delta\mu)(t-t')} \\
&\quad \times e^{-i(\delta\mu t - \varphi_{LR} - \mathbf{k}\cdot\mathbf{d})} f_R(\mathbf{k})^* f_L(\mathbf{k}), \qquad (5.42b)
\end{aligned}$$

where

$$\varphi_{LR} = \varphi_L - \varphi_R \qquad (5.43)$$

is the relative phase of the macroscopic wavefunctions and $\mathbf{d}$ is defined in Eq. (5.1).

We will show that the contributions $F_{L\to R}(t)$ ($F_{R\to L}(t)$) can be interpreted to describe the processes where atoms are outcoupled from the left to the right (right to the left) system. For this purpose we use Eq. (5.25b) in Eq. (5.22) for the propagator of the outcoupled atoms and obtain

$$\begin{aligned} F_{L\to R}(t) &= \Gamma \text{Re}\, e^{i(\delta\mu t - \varphi_{LR})} \int_0^t d\tau \int d\mathbf{r}\, d\mathbf{r}'\, e^{i(\mathbf{q}\cdot(\mathbf{r}-\mathbf{r}')-\Omega\tau)} \\ &\quad \times \left(\frac{im}{2\pi\hbar\tau}\right)^{\frac{3}{2}} \exp\left[-\frac{i}{\hbar}\frac{m(\mathbf{r}'-\mathbf{r})^2}{2\tau}\right] f_L(\mathbf{r}'-\mathbf{r}_L) f_R(\mathbf{r}-\mathbf{r}_R), \quad (5.44\text{a}) \\ F_{R\to L}(t) &= \Gamma \text{Re}\, e^{-i(\delta\mu t - \varphi_{LR})} \int_0^t d\tau \int d\mathbf{r}\, d\mathbf{r}''\, e^{i(\mathbf{q}\cdot(\mathbf{r}-\mathbf{r}')-(\Omega-\delta\mu)\tau)} \\ &\quad \times \left(\frac{im}{2\pi\hbar\tau}\right)^{\frac{3}{2}} \exp\left[-\frac{i}{\hbar}\frac{m(\mathbf{r}'-\mathbf{r})^2}{2\tau}\right] f_R(\mathbf{r}'-\mathbf{r}_R) f_L(\mathbf{r}-\mathbf{r}_L). \quad (5.44\text{b}) \end{aligned}$$

With the change of variables

$$\bar{\mathbf{r}} = \mathbf{r} - \mathbf{r}', \quad (5.45\text{a})$$

$$\mathbf{R} = \frac{\mathbf{r}+\mathbf{r}'}{2}, \quad (5.45\text{b})$$

Eq. (5.44) is rewritten as

$$\begin{aligned} F_{L\to R}(t) &= \Gamma \text{Re}\, e^{i(\delta\mu t - \varphi_{LR})} \int_0^t d\tau \left(\frac{im}{2\pi\hbar\tau}\right)^{\frac{3}{2}} e^{i(\omega\mathbf{q}-\Omega)\tau} \int d\bar{\mathbf{r}}\, d\mathbf{R} \\ &\quad \times \exp\left[\frac{i}{\hbar}\frac{m}{2\tau}\left(\bar{\mathbf{r}} - \frac{\hbar\mathbf{q}}{m}\tau\right)^2\right] f_L(\mathbf{R}-\frac{\bar{\mathbf{r}}}{2}+\mathbf{d}) f_R(\mathbf{R}+\frac{\bar{\mathbf{r}}}{2}), \quad (5.46\text{a}) \\ F_{R\to L}(t) &= \Gamma \text{Re}\, e^{-i(\delta\mu t - \varphi_{LR})} \int_0^t d\tau \left(\frac{im}{2\pi\hbar\tau}\right)^{\frac{3}{2}} e^{i(\omega\mathbf{q}-\Omega+\delta\mu)\tau} \int d\bar{\mathbf{r}}\, d\mathbf{R} \\ &\quad \times \exp\left[\frac{i}{\hbar}\frac{m}{2\tau}\left(\bar{\mathbf{r}} - \frac{\hbar\mathbf{q}}{m}\tau\right)^2\right] f_R(\mathbf{R}-\frac{\bar{\mathbf{r}}}{2}+\mathbf{d}) f_L(\mathbf{R}+\frac{\bar{\mathbf{r}}}{2}), \quad (5.46\text{b}) \end{aligned}$$

where $\mathbf{d}$ is defined in Eq. (5.1). The exponential in the integral over $\bar{\mathbf{r}}$ oscillates very fast with respect to the wave functions $f_j(\mathbf{r})$. Therefore the main contribution to the integral over $\bar{\mathbf{r}}$ comes from $\bar{\mathbf{r}}_0 = \frac{\hbar\mathbf{q}}{m}\tau$, where the term in the exponential vanishes. By means of the saddle point approximation we take the wave functions at the point $\bar{\mathbf{r}}_0$ out of the integral. We then perform the $\bar{\mathbf{r}}$ integration using the Fresnel integral [49]

$$\int_{-\infty}^{\infty} dt\, e^{i\gamma t^2} = \sqrt{\frac{\pi}{|\gamma|}} e^{i\,\text{sign}(\gamma)\pi/4} \qquad (5.47)$$

5.2 Scattered light intensity

and obtain

$$F_{L\to R}(t) \approx \Gamma \mathrm{Re}\, e^{i(\delta\mu t - \varphi_{LR})} \int_0^t d\tau\, e^{i(\omega \mathbf{q} - \Omega)\tau} \int d\mathbf{r} f_L\left(\mathbf{r} + \mathbf{d} - v_\mathbf{q}\tau\right) f_R(\mathbf{r}), \quad (5.48a)$$

$$F_{R\to L}(t) \approx \Gamma \mathrm{Re}\, e^{-i(\delta\mu t - \varphi_{LR})} \int_0^t d\tau\, e^{i(\omega \mathbf{q} - \Omega + \delta\mu)\tau} \int d\mathbf{r}\, f_L(\mathbf{r}) f_R\left(\mathbf{r} - \mathbf{d} - v_\mathbf{q}\tau\right), \quad (5.48b)$$

where $v_\mathbf{q} = \frac{\hbar \mathbf{q}}{m}$ is the recoil velocity. The approximation that leads to Eq. (5.48) is exact in the limit of homogeneous atomic systems. In such case the momentum distribution of the condensate fraction will be a delta function at zero momentum and all outcoupled atoms have exactly the same momentum $\hbar \mathbf{q}$ after the outcoupling process. The condensates we consider are confined by an external potential and have a finite extension in space which leads to a certain width δp in their momentum distribution. This leads to a spread in momentum around $\hbar \mathbf{q}$ for the outcoupled atoms. By taking the value of the atomic wave functions at the point of stationary phase of the exponential in Eq. (5.46) one neglects this spread in momentum. The saddle point approximation is thus applicable if $\hbar|\mathbf{q}| \gg \delta p$.

For completeness we also give the contributions of the macroscopic wavefunctions to the background current Eq. (5.29), making the same approximations as in Eqs. (5.48). Using Eq. (5.25b) in Eqs. (5.22) and one finds

$$F_L(t) \approx \Gamma \mathrm{Re} \int_0^t d\tau\, e^{i(\omega \mathbf{q} - \Omega)\tau} \int d\mathbf{r} f_L(\mathbf{r}) f_L(\mathbf{r} - v_\mathbf{q}\tau), \quad (5.49a)$$

$$F_R(t) \approx \Gamma \mathrm{Re} \int_0^t d\tau\, e^{i(\omega \mathbf{q} - \Omega + \delta\mu)\tau} \int d\mathbf{r}\, f_R(\mathbf{r}) f_R(\mathbf{r} - v_\mathbf{q}\tau). \quad (5.49b)$$

In the following we will assume that $\mathbf{d}$ and $\mathbf{q}$ point along the x axis. We see from Eqs. (5.48) that $F_{L\to R}(t)$ has only a contribution if the momentum of the outcoupled atoms $\hbar \mathbf{q}$ points into the same direction as $\mathbf{d}$, that is from left to right system. For the term $F_{R\to L}(t)$ this is just opposite and we can only get a contribution if $\mathbf{q}$ points along the negative x-direction. We also note that for small times $t < \frac{d_0}{v_q}$ where $d_0 = d - (\xi_L + \xi_R)/2$ is the effective distance of the two systems, the interference contribution vanishes $F_I(t) = 0$. Only if the outcoupled atoms from e.g. the left trap had enough time to travel to the right trap the interference contribution is nonzero[3]. It is tempting to interpret this fact as due to the loss of which way information [128, 129]. While for small times $t < \frac{d_0}{v_q}$ one could in principle determine which system scattered a photon by looking at the outcoupled atoms, this information gets lost once the atoms had enough time to propagate from the left to the right system. In general the which way

---

[3]For small times $t \to 0$ the function $\left(\frac{im}{2\pi\hbar\tau}\right)^{\frac{3}{2}} \exp\left[-\frac{i}{\hbar}\frac{m(\mathbf{r}'-\mathbf{r})^2}{2\tau}\right] \to \delta(\mathbf{r} - \mathbf{r}')$ in Eqs. (5.44) becomes a delta function in space. Thus we note that there is no interference contribution for small times, since the two systems have no spatial overlap $\int d\mathbf{r} f_L(\mathbf{r}) f_R(\mathbf{r}) = 0$.

information is lost as soon as the wavefunctions of the outcoupled atoms from both systems overlap.

However this intuitive picture is not entirely correct. It is important to note that the interference contribution Eqs. (5.48) is different from zero only when the outcoupled atoms from one system overlap with the wave function of the trapped atoms of the other system. Otherwise the contribution of the integral over the wave functions in Eqs. (5.48) vanishes. The interference contribution for the detected intensity arises from the fact that the outcoupled atoms from one system can be transferred into the other system by a second Raman transition. The loss of the which way information is hence not enough but the exchange of atoms from one to the other system is essential. It can be seen as Josephson coupling between the two spatially separated macroscopic wave functions linked via the outcoupled atoms [130].

5.3 A related experiment

The physical system we have considered in the previous sections is closely related to the experimental setup used by Saba *et al.* [47]. In this experiment two Bose-Einstein condensates were trapped in a double well potential with an energy offset ΔV between the two potential wells, due to a magnetic field gradient, indicated in Fig. (5.2a). The atoms were illuminated with two counterpropagating Bragg beams, where absorption of a photon from one beam and stimulated emission into the other beam leads to momentum transfer to some of the atoms. The Bragg-scattered atoms were outcoupled of the trap since the recoil energy was much higher than the trap depth.

After the Bragg beams were applied an image of the atoms was taken by illuminating them with a resonant laser beam. The atomic density distribution measured in this way is shown in Fig. (5.3a). Each panel displays an absorption image of the atoms for a different value of ΔV. The outcoupled atoms form an interference pattern [47]. The thick elongated shadows on the far left of each panel are the remaining Bose-Einstein condensates in the trap which were not spatially resolved.

Figure (5.3b) (taken from [47]) displays the measured light intensity as a function of time of the second Bragg beam for different values of the trap offset ΔV. The light signal exhibits an oscillatory behaviour, where the oscillation frequency changes with different values of the trap offset $\Delta V/\hbar$. The oscillations in the light intensity start only after a certain time t_c, which is in agreement with the estimate $t_c \approx d/v_q$, where d is the distance between the two condensates and v_q is the recoil velocity of the outcoupled atoms. A measurement of the oscillations hints to a frequency determined by the trap offset $\Delta V/\hbar$.

We use the model introduced in Sec. 5.1 in order to describe the experimental setup, thereby making several assumptions. In [47] the atoms were outcoupled of the trap due to the large momentum transfer and did not change their internal state during the outcoupling process. By approximating the outcoupling of the atoms by a Raman

5.3 A related experiment

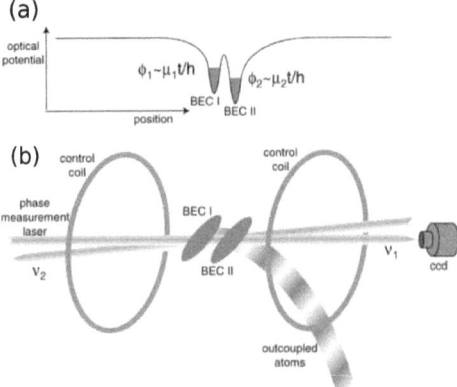

Figure 5.2: Experimental setup of the MIT experiment. (a) energy diagram of the two Bose-Einstein condensates trapped in a double well potential. (b) experimental scheme for monitoring the outcoupled atoms by continuously measuring the light intensity of one of the laser beams. Figure is taken from [47].

transition to an untrapped state one neglects (i) the additional mean-field potential, due to the condensed atoms, which the outcoupled atoms feel and (ii) the tunneling barrier the atoms have to overcome in order to leave the trap.

The mean field energy was about 2% of the kinetic energy $E_{\rm kin}$ of the outcoupled atoms, while $E_{\rm kin}$ was about 5 times larger than the trap depth [47, 131]. Hence the effect of the trap and of the additional mean field potential constitute only small perturbations to the outcoupling process and we can apply the equations derived in Sec. (5.2) for the scattered photons to the experimental setup[4].

While real experiments always operate at some finite temperature T we assume $T = 0$ for simplicity and neglect the effect of the noncondensed atoms [5]. The trapping potentials for the two condensates are given by

$$V_L(\mathbf{r}) = V(\mathbf{r}) + \Delta V, \quad (5.50a)$$
$$V_R(\mathbf{r}) = V(\mathbf{r}) \quad (5.50b)$$

---

[4]The experimentally measured oscillation frequency of the light signal shown in Fig. (5.3b) is on the order of a few kHz, while the integration time of a photo detector is on the order of $10^{-9} s$. Thus we see that the approximation made by taking the photon flux out of the integral in Eq. (5.13) is well justified for the experimental parameters of [47].

[5]The amount of noncondensed atoms compared to the condensed fraction is negligible for the experimental parameters [47].

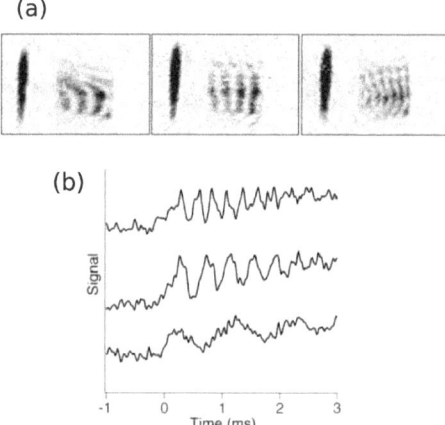

Figure 5.3: Experimental results taken from [47]. (a) Time of flight image of the atoms after applying the Bragg beams for different values of the energy offset ΔV between the two potential minima. (b) Measured light intensity of the second Bragg beam as a function of time for different values of ΔV. Figure is taken from [47]

5.3 A related experiment

with

$$V(\mathbf{r}) = \frac{1}{2}m(\omega_x^2 x^2 + \omega_y^2 y^2 + \omega_z^2 z^2)$$

$$= \mu \sum_{j=(x,y,z)} \frac{j^2}{r_j^2}, \qquad (5.51)$$

where μ is the chemical potential, $r_j = \sqrt{\frac{2\mu}{m\omega_j^2}}$ and the radial size of the wavefunctions is given by $\xi_L = \xi_R = \bar{\xi} = 2r_x$. The condensate wave functions using the Thomas-Fermi approximation read

$$f_j(\mathbf{r}) = [\frac{1}{g}(\mu - V(\mathbf{r}))]^{1/2}, \qquad (5.52)$$

and μ is determined by the equation $N_C = \int d\mathbf{r} |f_j(\mathbf{r})|^2$, where N_C is the number of atoms in each condensate. The chemical potential is the same for both systems, since we assume equal atom numbers N_C for both condensates according to the experimental parameters [47] and is given by [122]

$$\mu = \frac{15^{2/5}}{2}\left(\frac{N_C a_s}{\bar{a}}\right)^{2/5} \hbar\bar{\omega}, \qquad (5.53)$$

where we introduced

$$\bar{a} = \sqrt{\frac{\hbar}{m\bar{\omega}}}, \qquad (5.54\text{a})$$

$$\bar{\omega} = (\omega_x \omega_y \omega_z)^{1/3}. \qquad (5.54\text{b})$$

The Thomas-Fermi approximation gives good results if $\frac{N_C a_s}{\bar{a}} \gg 1$, where a_s is the s-wave scattering length of the atoms [122]. For the experimental parameters of [47] one finds $\frac{N_C a_s}{\bar{a}} \approx 560$, such that the application of the Thomas-Fermi approximation for the condensate wavefunctions is well justified [61]. In the present case the difference in chemical potential introduced in Eq. (5.23) is given by the offset ΔV of the trapping potentials

$$\delta\mu = \frac{\mu_L - \mu_R}{\hbar} = \Delta V/\hbar. \qquad (5.55)$$

Using Eq. (5.52) for the condensate wave functions in Eqs. (5.48) and Eqs. (5.49) the photon flux for outcoupling to the right (**q** points along the positive x-direction) reads

$$F_C(t) \approx 2\pi\Gamma N_C K_m \left(1 + \frac{K(\omega_q - \Omega)}{K_m}\cos\left[\delta\mu t - \varphi_{LR} + (\omega_\mathbf{q} - \Omega)\frac{d}{v_q}\right]\Theta\left(t - \frac{d - 2r_x}{v_q}\right)\right). \qquad (5.56)$$

Here the Heaviside function takes into account that there is no contribution from the interference terms if the outcoupled atoms had not enough time to cross the distance between the two condensates as discussed below Eq. (5.49), and $K_m = \frac{1}{2}(K(\omega_q - \Omega) + K(\omega_q - \Omega + \delta\mu))$ with

$$K(x) = \sqrt{\frac{t_0^2}{5\pi}} e^{-\frac{t_0^2}{5}x^2}, \quad (5.57)$$

where $t_0 = \frac{r_x}{v_q}$ and r_x is the Thomas-Fermi radius of the condensates in x-direction. The detailed steps of the derivation of Eq. (5.56) are given in Appendix (C). It describes an oscillatory behaviour of the photon flux where the oscillation frequency is given by $\delta\mu$.

Applying a second pair of Bragg beams to the setup shown in Fig. (5.2) one can simultaneously outcouple atoms to the right and to the left [130]. In Eq. (C.10) we see that the two contributions $F_{L \to R}$ and $F_{R \to L}$ have an offset in phase Θ which is given by

$$\Theta = \frac{d}{v_q}(2\Omega - 2\omega_q - \delta\mu). \quad (5.58)$$

Eq. (5.58) agrees with the corresponding expression in [130] derived from a phenomenological model taking into account the difference in notation ($q = 2k_r$). As shown in Eq. (5.38) measuring the photon flux or the flux of outcoupled atoms is equivalent and we can interpret Θ as a phase shift between the left and right outcoupled atom fluxes. In [130] it is shown that this phase shift can be interpreted as the phase the atoms accumulate when moving from one condensate to the other. Due to the dependence on the detuning between the two Bragg beams ω_{12} it can be controlled experimentally. Its rate of change as a function of the Bragg-frequency ω_{12} is given by

$$\frac{d\Theta}{d\Omega} = 2\frac{d}{v_q}, \quad (5.59)$$

thus by measuring Θ as a function of ω_{12} one finds a linear behaviour with slope $2\frac{d}{v_q}$ [130].

We now use Eq. (5.30) and Eq. (5.42) to calculate the condensate contribution of the photon flux for the present setup and compare the results obtained with the approximate expressions Eq. (C.7). The momentum distribution of a Bose Einstein condensate in the Thomas-Fermi approximation in a harmonic trap can be obtained by Fourier transformation of the wave function Eq. (5.52) and is given by [122]

$$|f(\mathbf{k})|^2 = \frac{15\pi^3 N_C \bar{R}}{2}\left(\frac{J_2(\mathbf{p})}{\mathbf{p}^2}\right)^2, \quad (5.60)$$

where $J_2(\mathbf{p})$ is the Bessel function of order 2 and $p^2 = k_x^2 r_x^2 + k_y^2 r_y^2 + k_z^2 r_z^2$. We choose the normalization of $f(\mathbf{k})$ such that $\int \frac{d\mathbf{k}}{(2\pi)^3}|f(\mathbf{k})|^2 = N_C$. Using Eq. (5.60) into Eq. (5.30)

5.3 A related experiment

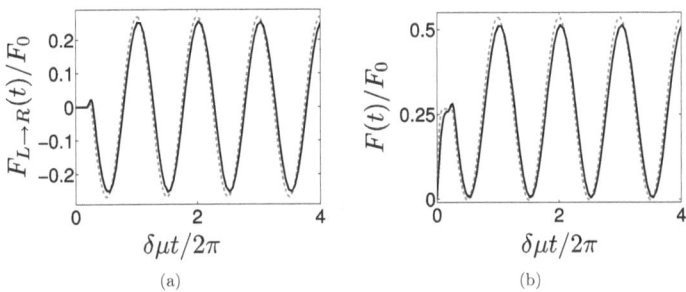

Figure 5.4: Photon fluxes computed by numerically integrating Eq. (5.61) (solid black line) compared to the approximate results obtained from Eq. (C.7) (dashed red line). a) Interference Flux $F_{L\to R}(t)$ in units of F_0 as a function of time in units of $\frac{\delta\mu}{2\pi}$. b) Total photon Flux $F(t)$ in units of F_0 as a function of time in units of $\frac{\delta\mu}{2\pi}$. Other parameters are $\omega_q = \Omega$, $r_x = 3\mu\text{m}$, $d = 5r_x$, $v_q = 6\frac{\text{cm}}{\text{s}}$, $\varphi_{LR} = 0$ and $\frac{\delta\mu}{2\pi} = 10^3\text{Hz}$.

and Eq. (5.42) we find for the photon fluxes

$$F_L(t) = \pi F_0 \int_1^1 dx \int_0^\infty dp \left(\frac{J_2(p)}{p}\right)^2 \delta^{t/2t_0}\left(A_\delta(\Omega - \omega_q)\right), \quad (5.61\text{a})$$

$$F_R(t) = \pi F_0 \int_1^1 dx \int_0^\infty dp \left(\frac{J_2(p)}{p}\right)^2 \delta^{t/2t_0}\left(A_\delta(\Omega - \omega_q - \delta\mu)\right), \quad (5.61\text{b})$$

$$F_{L\to R}(t) = 2\pi F_0 \int_1^1 dx \int_0^\infty dp \left(\frac{J_2(p)}{p}\right)^2 \delta^{t/2t_0}\left(A_\delta(\Omega - \omega_q)\right)$$
$$\times \cos\left[px\frac{d\cos\alpha}{r_x} + \varphi_{LR} + \delta\mu t + A_\delta(\Omega - \omega_q)\frac{t}{2t_0}\right]$$
$$\times J_0\left(p\sqrt{1-x^2}\frac{d\sin\alpha}{r_x}\right), \quad (5.61\text{c})$$

$$F_{R\to L}(t) = 2\pi F_0 \int_1^1 dx \int_0^\infty dp \left(\frac{J_2(p)}{p}\right)^2 \delta^{t/2t_0}\left(A_\delta(\Omega - \omega_q - \delta\mu)\right)$$
$$\times \cos\left[px\frac{d\cos\alpha}{r_x} + \varphi_{LR} + \delta\mu t - A_\delta(\Omega - \omega_q - \delta\mu)\frac{t}{2t_0}\right]$$
$$\times J_0\left(p\sqrt{1-x^2}\frac{d\sin\alpha}{r_x}\right), \quad (5.61\text{d})$$

with F_0 given in Eq. (C.8), $J_j(x)$ is the Bessel function of order j and

$$A_\delta(\Omega) = \Omega t_0 - \frac{1}{2q'}(p^2 + 2pq'x). \qquad (5.62)$$

In deriving Eqs. (5.61) we assumed outcoupling to take place only in the $x-y$ plane. We took the z- axis of the integration along the $\mathbf{q}$ direction, such that α is the angle between $\mathbf{d}$ and $\mathbf{q}$ as indicated in Fig. (5.5). Equations (5.61) has been obtained within

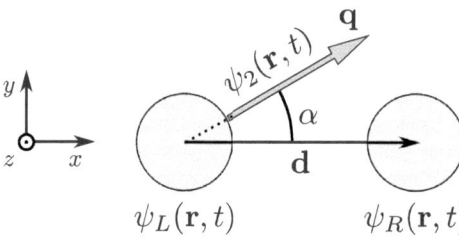

Figure 5.5: Sketch of outcoupling of atoms from two condensates into different directions, where the direction of view is along the long axis of the condensates. The field operator $\psi_{L,R}(\mathbf{r},t)$ describes the atoms in the left (right) trap and $\psi_2(\mathbf{r},t)$ the freely propagating atoms. The coordinate system to the left serves for orientation and α is the angle between the outcoupling direction $\mathbf{q}$ and the distance vector $\mathbf{d}$

our perturbative treatment without any approximations but it does not allow for a clear physical interpretation of the different contributions. This is a clear disadvantage compared with the approximate expressions Eqs. (C.7). In Fig. (5.4) we show the interference contribution $F_{L \to R}(t)$ and the total photon flux $F(t)$ calculated from Eq. (5.61) for $\alpha = 0$ in comparison with the approximate results obtained from Eq. (C.7). We get very good quantitative agreement between the two results, showing that the saddle point approximation made in Eq. (5.48) yields good results for the experimental parameters.

Comparing Fig. (5.4) with the experimental results of [47] for the laser intensity shown in Fig. (5.3) we find that our theory predicts correctly the oscillation period and also accounts for the fact why the interference current needs some time to build up. We note that the linear response treatment does not predict any decrease in the visibility of the interference pattern with time, in marked contrast to the experimental results. Possible reasons for the decrease in visibility in the experimental data are heating of the condensate, depletion and spontaneous Rayleigh scattering [47]. All these effects are not taken into account in our linear response treatment.

Figure (5.6a) displays the interference flux $F_{L \to R}(t)$ as a function of time t and angle α. We note that the interference contribution vanishes rapidly as we change the outcoupling angle α from 0 (outcoupling to the right) to higher values.

5.4 Measuring the atomic field

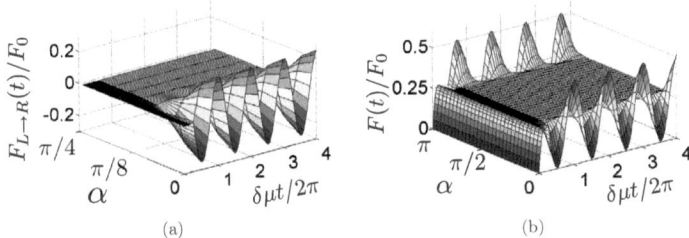

Figure 5.6: Photon fluxes computed by numerically integrating Eq. (5.61). a) Interference Flux $F_{L\to R}(t)$ in units of F_0 as a function of time in units of $\frac{\delta\mu}{2\pi}$ and α. b) Total interference flux $F(t)$ in units of F_0 as a function of time in units of $\frac{\delta\mu}{2\pi}$ and α. Other parameters are as in Fig. (5.4)

In Fig. (5.6b) we show the total photon flux $F(t)$ as a function of time and angle α. It oscillates in time for $\alpha \approx 0$ and $\alpha \approx \pi$. These values of α exactly corresponds to outcoupling to the right and left respectively, as seen in Fig. (5.5). For angles $\frac{\pi}{8} < \alpha < \frac{7\pi}{8}$ there is a plateau in the photon flux and the only contributions to the photon flux are the background contributions $F_L(t)$ and $F_R(t)$.

From Eq. (5.48) one expects a contribution from $F_{L\to R}(t)$ provided that the displaced wave function of the left condensate $f(\mathbf{r}+\mathbf{d}-t\frac{\hbar}{m}\mathbf{q})$ will overlap with the wave function of the right condensate for some values of t. From a simple geometric argument one then expects that there should be a contribution up to angles $\alpha = \arctan\frac{2r_r}{d}$. This corresponds for our parameters to $\alpha \approx \pi/8$, which agrees well with the calculated result, as seen in Fig. (5.6a). The fact that the oscillations in the scattered light intensity vanish for $\frac{\pi}{8} < \alpha < \frac{7\pi}{8}$ shows that they cannot be due to interference of the outcoupled atoms from the two condensates, since for long times the wavefunctions of the outcoupled atoms from both systems will overlap also at these angles. Thus interpreting the oscillations in the light intensity as consequence of the beating of two atom lasers [47] is not valid. Instead the interference contribution to the photon flux is due to a Josephson type coupling of the two condensates mediated by the outcoupled atoms, as discussed at the end of Sec. (5.2.2).

5.4 Measuring the atomic field

We now show how the setup discussed in Sec. (5.1) can be used to measure the expectation value of the atomic field operator $\langle\psi(\mathbf{r},t)\rangle$ for an ultracold bosonic system. Such a measurement corresponds in optics to determine the quadrature of the field by means of a homodyne detection, where the field to detect is superposed at a beam-splitter with a local oscillator, namely, a laser field [132]. We first review some properties of optical homodyne detection in order to point out the similarities and differences between optical and atomic homodyne detection.

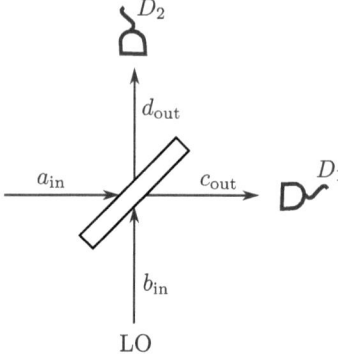

Figure 5.7: Schematic diagram for optical homodyne detection. The field to measure is described by the photon operator a_in and the local oscillator field by b_in. They are mixed at a 50-50-beam-splitter and the output fields described by the photon operators c_out and d_out are measured at the detector D_1 and D_2 respectively.

Figure (5.7) shows a schematic setup for an optical homodyne measurement. The photon operators for the input fields a_in and b_in are related to the output photon operators c_out and d_out via the beam-splitter relation

$$c_\text{out} = \frac{1}{\sqrt{2}}(a_\text{in} + \imath b_\text{in}), \qquad (5.63\text{a})$$

$$d_\text{out} = \frac{1}{\sqrt{2}}(\imath a_\text{in} + b_\text{in}), \qquad (5.63\text{b})$$

where we assumed a 50-50-beam-splitter. The signals measured by the two detectors are determined by the operators

$$c_\text{out}^\dagger c_\text{out} = \frac{1}{2}\left(a_\text{in}^\dagger a_\text{in} + b_\text{in}^\dagger b_\text{in} + \imath(a_\text{in}^\dagger b_\text{in} + b_\text{in}^\dagger a_\text{in})\right), \qquad (5.64\text{a})$$

$$d_\text{out}^\dagger d_\text{out} = \frac{1}{2}\left(a_\text{in}^\dagger a_\text{in} + b_\text{in}^\dagger b_\text{in} - \imath(a_\text{in}^\dagger b_\text{in} + b_\text{in}^\dagger a_\text{in})\right). \qquad (5.64\text{b})$$

We assume the field of the local oscillator described by b_in to be in a coherent state $|\beta_L\rangle$ with well defined and controllable phase Φ_L and the frequency of the local oscillator to be equal to the frequency of the input field. An ordinary optical homodyne detection

5.4 Measuring the atomic field

consists in measuring the light intensity at detector D_1, given by [132]

$$I_1(\Phi_L) \propto \langle c_{\text{out}}^\dagger c_{\text{out}} \rangle = \frac{N_{\text{tot}}}{2}(1 + V_O \sin(\Phi_L - \phi_a)) \tag{5.65}$$

where $N_{\text{tot}} = \langle a_{\text{in}}^\dagger a_{\text{in}} \rangle + |\beta_L|^2$ is the total average number of photons of the input fields and ϕ_a is the phase of the input field described by a_{in}, defined by $\langle a_{\text{in}} \rangle = |\langle a_{\text{in}} \rangle| e^{i\phi_a}$. The quantity V_O is the visibility of the measured light intensity is given by

$$V_O = \frac{2|\beta_L||\langle a_{\text{in}} \rangle|}{|\beta_L|^2 + \langle a_{\text{in}}^\dagger a_{\text{in}} \rangle}. \tag{5.66}$$

The visibility V_O quantifies the contrast of the interference signal of the scattered light when measured as a function of the phase Φ_L. In the case of a strong local oscillator $|\beta_L|^2 \gg \langle a_{\text{in}}^\dagger a_{\text{in}} \rangle$ one finds from Eq. (5.66)

$$|\langle a_{\text{in}} \rangle| \approx |\beta_L| V_O / 2 \,. \tag{5.67}$$

We note that by measuring the visibility V_O one can determine the mean value of the photonic field operator $|\langle a_{\text{in}} \rangle|$.

A different method to determine the mean value of the photonic field operator $|\langle a_{\text{in}} \rangle|$ is the so called balanced homodyne detection. In this case one makes also use of the second detector D_2. We note from Eqs. (5.64) that the background signal to the light intensities have the same sign, while the interference contributions have opposite sign. The difference between the light intensities measured at the two detectors leads to

$$I_{\text{diff}} \propto \langle c_{\text{out}}^\dagger c_{\text{out}} - d_{\text{out}}^\dagger d_{\text{out}} \rangle = |\langle a_{\text{in}} \rangle||\beta_L| \sin(\Phi_L - \phi_a) \,. \tag{5.68}$$

In this case the measured signal is directly proportional to $|\langle a_{\text{in}} \rangle|$.

We now compare the setup discussed in Sec. (5.1) with optical homodyne detection. Measurement of the mean value of a field operator can be performed for atoms by using the setup shown in Fig. (5.1), where we assume the left system to be a Bose-Einstein condensate with known order parameter serving as local oscillator. Indeed Eq. (5.22b) shows that the interference contribution to the photon flux is proportional to the product of the atomic field operators in close analogue to the interference terms in Eqs. (5.64) for the optical setup. In analogy to optical homodyne detection we can identify the Raman laser shown in Fig. (5.1) as beam-splitter for the atoms. However the Raman lasers perform an indirect beam-splitter transformation for the atoms, where the outcoupled atoms in state $|2\rangle$ serve as intermediate state. In order to mix the atomic fields, the outcoupled atoms from one system must be transferred to the other system via the freely propagating atoms in state $|2\rangle$. This yields an intuitive explanation why the interference contribution of the photon flux Eq. (5.22) cannot be due to interference of the outcoupled atoms from both systems, as discussed at the end of Sec. (5.1).

Equation (5.67) shows that for an ordinary optical homodyne detection the visibility V_O can be used to determine $|\langle a_{\rm in} \rangle|$. For the atomic homodyne detection we define the visibility of the photon flux Eq. (5.15) as

$$V_A = \frac{\max(F(t)) - \min(F(t))}{\max(F(t)) + \min(F(t))}, \qquad (5.69)$$

where the minimum and maximum is determined in a time interval which is much larger than the oscillation frequency of the interference contribution $F_I(t)$. Thereby $F(t)$ is taken as a function of time only, keeping all other parameters fixed. The visibility V_A of the scattered light intensity, defined in Eq. (5.69), in general behaves as

$$V_A \approx \frac{2|\langle\psi_L\rangle\langle\psi_R\rangle|}{|\langle\psi_L\rangle|^2 + |\langle\psi_R\rangle|^2 + C_Q}, \qquad (5.70)$$

where C_Q is a constant due to the contribution of noncondensed atoms Eqs. (5.31). Comparing Eq. (5.70) with Eq. (5.66) we note that they have the same form up to the constant C_Q. Thus for a strong local oscillator, corresponding to the condition $|\langle\psi_L\rangle|^2 \gg |\langle\psi_R\rangle|^2 + C_Q$, we see that measurement of the visibility of the photon flux $F(t)$ can be used to determine the order parameter of the right system similar to Eq. (5.67) for the optical setup. There is however a crucial difference between the optical setup and the atomic homodyne setup discussed here. Equation (5.70) describes the visibility of the measured light intensity that yields information about the atomic gas, in contrast to optical homodyne measurements in which one measures the optical fields directly.

In the following we will consider extensions of the experimental setup of [47] and show how they may allow one for measuring the temperature of a Bose-Einstein condensate or monitor the superfluid Mott-insulator transition by photon counting. Similar to balanced homodyne detection in optics, we will consider a scheme which allows one to get rid of the background contributions of the photon flux. Therefore we define the time average of a function $O(t)$ as

$$\langle O(t) \rangle_t = \frac{1}{\Delta T} \int_{t_c}^{t_c + \Delta T} F_I(t), \qquad (5.71)$$

where t_c is the time the atoms need to cross the distance between the left and right system and ΔT is a time interval large compared to the inverse oscillation frequency of $F_I(t)$,

$$\Delta T \gg \frac{\hbar}{\mu_L - \mu_R}. \qquad (5.72)$$

From Eq. (5.42) one can show that the time average of the interference contribution vanishes $\langle F_I(t) \rangle_t = 0$ and thus Eq. (5.15) leads to

$$F_I(t) = F(t) - \langle F(t) \rangle_t. \qquad (5.73)$$

5.4 Measuring the atomic field

Equation (5.73) shows that the measured photon flux $F(t)$ also permits to determine the interference term $F_I(t) \propto \langle\psi_L\rangle\langle\psi_r\rangle$, by subtracting the time average $\langle F(t)\rangle_t$ from the measured signal. Knowledge of $\langle\psi_L\rangle$ then allows one to determine the mean value of the atomic field operator of the right system $\langle\psi_R\rangle$ by photon counting.

5.4.1 Temperature Measurement of a Bose-Einstein condensate

In this section we will show how an extension of the scheme studied in Sec.(5.3) can be used to measure the temperature of a Bose-Einstein condensate. We consider two Bose-Einstein condensates confined at two separate regions in space as shown in Fig. (5.1), where the trapping potentials for the two condensates are given by Eqs. (5.50). The left condensate serves as our reference system. We assume it to be at some fixed known reference temperature T_{ref}. For simplicity we will assume our reference condensate to be at zero temperature $T_{\text{ref}} = 0$. We will see that this condition is not necessary in order to determine the temperature of the right condensate. The right Bose-Einstein condensate is at some finite temperature T which we want to specify by measuring the scattered light from the system.

We now calculate the interference contribution $F_I(t)$ for the present setup and show that its measurement may allow one to determine the temperature of the right condensate. We assume outcoupling to the right (**q** points along the positive x-axis), such that $F_{L\to R}(t)$ is the only contribution to interference photon flux $F_I(t)$. From Eq. (5.44) we note that

$$F_{L\to R}(t) \propto \sqrt{N_C}, \qquad (5.74)$$

where N_C is the number of condensed atoms in the right trap, which depends on the temperature T and is given by [61]

$$N_C(T) = N\left(1 - t_r^3 - \frac{\zeta(2)}{\zeta(3)} t_r^2 \eta (1 - t_r^3)^{2/5}\right). \qquad (5.75)$$

The parameter $t_r = \frac{T}{T_c}$ is the reduced temperature and T_c is the critical temperature for a noninteracting Bose gas in a harmonic trap, given by [61]

$$kT_c = \hbar\bar{\omega}\left(\frac{N}{\zeta(3)}\right)^{1/3} \approx 0.94\hbar\bar{\omega} N^{1/3}, \qquad (5.76)$$

where N is the total number of atoms in the trap and $\zeta(x) = \sum_{k=1}^{\infty} \frac{1}{k^x}$ is the Rieman Zeta-function. A further quantity which is needed in order to calculate $F_{L\to R}(t)$ is the chemical potential μ of the right condensate, which also depends on temperature in the present case and is given by [61]

$$\begin{aligned}\mu_R &= \frac{15^{2/5}}{2}\left(\frac{N_C a_s}{\bar{a}}\right)^{2/5} \hbar\bar{\omega} \\ &= \eta kT_c(1 - t_r^3)^{2/5}. \end{aligned} \qquad (5.77)$$

In Eq. (5.75) and Eq. (5.77) we have introduced the parameter η, defined as

$$\eta = \frac{\mu(T=0)}{kT_c}$$
$$\approx 1.57 \left(N^{1/6}\frac{a_s}{\bar{a}}\right)^{2/5}. \qquad (5.78)$$

It is the ratio between the chemical potential at zero temperature $T=0$ obtained from Eq. (5.77) and the critical temperature defined in Eq. (5.76). The parameter η can be used to discuss the effects of interactions on the thermodynamic behaviour of the Bose-Einstein condensate and one can show that in the thermodynamic limit the system exhibits a scaling behaviour in this parameter [61].

Using Eq. (5.75) and Eq. (5.77) in Eq. (5.61c) the interference contribution $F_{L\to R}(t)$ at finite temperature reads

$$F_{L\to R}(t) = 2\pi F_T(t_r) \int_1^1 dx \int_0^\infty dp \left(\frac{J_2(p)}{p}\right)^2 \qquad (5.79)$$
$$\times \delta^{t/2t_0}\left(A_\delta(\Omega-\omega_q)\right) J_0\left(p\sqrt{1-x^2}\frac{d\sin\alpha}{r_x}\right)$$
$$\times \cos\left[px\frac{d\cos\alpha}{r_x} + \varphi_{LR} + \delta\mu t + A_\delta(\Omega-\omega_q)\frac{t}{2t_0}\right],$$

where $A_\delta(\Omega)$ is defined in Eq. (5.62) The interference flux $F_{L\to R}(t)$ oscillates in time at frequency

$$\delta\mu = \delta\mu_0 + 0.94 N^{1/3}\bar{\omega}\eta\left[1-(1-t_r^3)^{2/5}\right] \qquad (5.80a)$$

and the amplitude of the oscillation is determined by

$$F_T(t_r) = \frac{15\pi\Gamma t_0 N}{8}\sqrt{1-t_r^3 - 1.37 t_r^2 \eta(1-t_r^3)^{2/5}}. \qquad (5.80b)$$

In Eq. (5.80a) the difference in chemical potential $\delta\mu_0 = \Delta V$ is due to the offset in the trap depth. We note that both, the frequency and the amplitude of $F_{L\to R}(t)$ are temperature dependent. In order to extract the temperature of the right condensate from the photon flux $F_{L\to R}(t)$ we introduce the quantity

$$A(T) = (\max(F_{L\to R}(t)) - \min(F_{L\to R}(t)))^2, \qquad (5.81)$$

where we take $F_{L\to R}(t)$ as a function of time t, for fixed temperature T and the minimum and maximum is taken over a time interval large compared to the inverse oscillation frequency of $F_{L\to R}(t)$. From Eq. (5.74) one finds that

$$A(T) = A(0) n_C(T), \qquad (5.82)$$

where $n_C(T) = N_C(T)/N$ is the condensate fraction in the right trap. We now evaluate the interference contribution $F_{L\to R}(t)$ by numerically integrating Eq. (5.79).

5.4 Measuring the atomic field

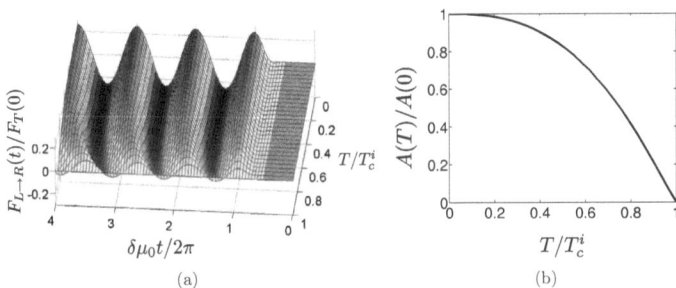

Figure 5.8: a) Interference contribution of the photon flux $F_{L\to R}(t)$ in units of $F_T(0)$ as a function of time in units of $\frac{2\pi}{\delta\mu}$ and temperature T in units of critical temperature T_c^i for an interacting Bose-Einstein condensate. It is determined from the condition $N_C(T) = 0$ in Eq. (5.75) and has to be evaluated numerically. b) $A(T)$, Eq. (5.81) in units of $A(0)$ is displayed as a function of temperature T in units of T_c^i. The condensates are made out of $N = 10^6$ sodium atoms with $\omega_{x,y} = 325$Hz radial trapping frequencies, $\omega_z = 10$Hz longitudinal trapping frequency and the scattering length $a_s = 55a_0$ with a_0 being the Bohr radius. We also use $v_q = 6\frac{\text{cm}}{\text{s}}$, $\delta\mu_0 = 2\pi 10^3$Hz, $d = 5r_x$, $\Omega - \omega_q = 0$, $\alpha = 0$ and $\varphi_{LR} = 0$.

Figure (5.8a) displays the interference contribution $F_{L\to R}(t)$ as a function of time and temperature T. Initially there is no interference current due to the finite time t_c the outcoupled atoms need to cross the distance between the two Bose-Einstein condensates. For times $t > t_c$ the interference signal for the photon flux builds up and shows oscillations with frequency $\delta\mu$. In the present case the amplitude of the interference fringes depends on the temperature T of the right condensate. One can also observe that the oscillation frequency $\delta\mu$ depends on T as shown in Eq. (5.80). By measuring such a change of the oscillation frequency in the scattered light signal one could hence verify the temperature dependence of the chemical potential, see Eq. (5.77). In practice this is very difficult since the chemical potential depends on the number of atoms in the condensate N_C. Due to the outcoupling of the atoms the condensates are depleted and hence N_C varies, which also leads to a change in the chemical potential. This effect is not taken into account in our perturbative treatment, since we assume that the change of atom number in the condensates due to the outcoupling is negligible compared to the total atom number in the condensates.

In Fig. (5.8b) we show $A(T)$ in units of $A(0)$ which is equal to the condensate fraction $n_C(T)$ in the right trap, see Eq. (5.82). At zero temperature $n_C(T) = 1$ corresponding to a pure condensate and for higher temperatures the condensate fraction decreases according to Eq. (5.75) up to the critical temperature T_C^i at which the condensate fraction vanishes. Here T_c^i is the critical temperature taking the interactions into account and is determined from the condition $N_C(T) = 0$ in Eq. (5.75). Extracting $A(T)$ from the measured light intensity and comparing with Eq. (5.75) hence may allow one to determine the temperature of the right condensate. We remark that Eq. (5.82) is independent on the temperature of the left condensate and thus its explicit knowledge is not necessary. However in order to calibrate the thermometer one should know $F_{L\to R}(t)$ at $T = 0$ for the right condensate in order to determine $A(0)$. We note that this setup requires to hold the atom numbers of the two condensates fixed for different experimental runs and also the temperature of the reference condensate. A quantitative measurement of the temperature will probably be difficult in practice, since depletion of the condensate due to the outcoupling process and additional Rayleigh scattering diminish the interference contribution [47].

5.4.2 Measurement of the superfluid order parameter

In this section we apply the scheme discussed in Sec.(5.3) to monitor the superfluid Mott-insulator transition for ultracold atoms in an optical lattice, by photon counting. We consider a setup similar to the one in the previous section, where we assume that the two atomic gases are tightly confined in the x-direction by a harmonic potential as shown in Fig. (5.9). We assume the trapping frequency ω_x in the x direction to be large enough $\hbar\omega_x \gg \mu, k_B T$ such that the atomic wavefunctions are in the ground state of the harmonic potential in this direction. In this case the motional degrees of freedom in the x-direction are frozen out and the systems are essentially two dimensional. For

5.4 Measuring the atomic field

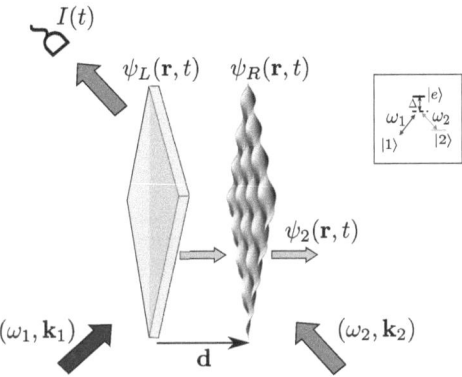

Figure 5.9: Setup for measuring the superfluid fraction in a 2D optical lattice. We consider a variation of Fig. (5.1), where the left system is assumed to be a two-dimensional homogeneous Bose-Einstein condensate which serves as our reference system and the right system is formed by ultracold atoms in an two-dimensional optical lattice. The atomic planes of the two systems are parallel to each other in the y-z plane shifted by the distance vector $\mathbf{d}$. Outcoupling is performed parallel to $\mathbf{d}$ such that the outcoupled atoms propagate along the positive x-direction as indicated by the green arrows. This setup could be realized in a 3D optical lattice making the lattice strength in the x-direction much stronger than in the y and z-directions such that one obtains an array of atomic planes. Then one removes all but two planes and switches off the lattice lasers in the y and z direction for the left of the remaining planes.

convenience we assume the left system to be a homogeneous two dimensional Bose-Einstein condensate[6] whereas the right system is confined by an optical lattice in the $y-z$ plan such that the potentials for the atoms are given by

$$V_L(\mathbf{r}) = \frac{1}{2}m\omega_x^2 x^2, \tag{5.83}$$

$$V_R(\mathbf{r}) = \frac{1}{2}m\omega_x^2 x^2 + V_0\left(\sin^2\frac{\pi y}{d_0} + \sin^2\frac{\pi z}{d_0}\right). \tag{5.84}$$

We expand the field operator for the right system into Wannier functions

$$\psi_R(\mathbf{r}) = f_0(x)\sum_l w_l(y,z) b_l, \tag{5.85}$$

where $f_0(x) = \frac{1}{\sqrt{\xi\sqrt{\pi}}}e^{-x^2/(2\xi^2)}$ is the motional ground state in the x-direction, $w_l(y,z)$ is the Wannier function centered at (y_{l_y}, z_{l_z}) and b_l annihilates an atom at lattice site l. Splitting the field operators according to Eq. (5.27) we find for the macroscopic wave functions

$$\Phi_L(\mathbf{r}) = f_0(x)\sqrt{n_L}e^{i\varphi_L}, \tag{5.86a}$$

which describes the Bose-Einstein condensate and n_L is the density of condensed atoms in the left trap. For the optical lattice the macroscopic wavefunction of the atoms reads

$$\Phi_R(\mathbf{r}) = f_0(x)\langle b\rangle \sum_l w_l(y,z), \tag{5.86b}$$

where $\langle b\rangle = \langle b_l\rangle$ is the superfluid order parameter, which is independent on the lattice site. The superfluid order parameter determines the quantum phase transition from superfluid to Mott-insulator phase, as discussed in Sec. (2.3). It is different from zero in the superfluid phase and vanishes in the Mott-insulator phase. We use the expansion of the field operator Eq. (5.85) in the Hamiltonian for the right system H_R, given in Eq. (5.6a). Following the steps of Sec. 2.3 we note that the dynamics of the right system is given by the Bose-Hubbard Hamiltonian Eq. (2.33) in 2 dimensions.

Using Eq. (5.86a) in Eq. (5.26a) and approximating the Wannier functions as Gaussian

$$w_l(y,z) = \frac{1}{w_0\sqrt{\pi}}e^{-\frac{(y-y_{l_y})^2 + (z-z_{l_z})^2}{2w_0^2}}, \tag{5.87}$$

we find for the background contribution to the photon flux of the condensate fraction

$$\frac{F_L(t)}{L^2} = \xi\pi^{1/2}n_L\Gamma\int dk_x e^{-k_x^2\xi^2}\delta^t(\Omega - \omega_{q+k_x}), \tag{5.88a}$$

$$\frac{F_R(t)}{L^2} = \frac{4\pi^{3/2}\xi w_0^2\langle b\rangle^2\Gamma}{d_0^4}\int dk_x \sum_{\mathbf{G}}\delta^t(\Omega - \omega_{q+k_x+\mathbf{G}})e^{-(k_x^2\xi^2 + \mathbf{G}^2 w_0^2)}. \tag{5.88b}$$

---

[6]Strictly speaking a homogeneous 2 dimensional Bose-Einstein condensate does not exist [133]. In real experiments a finite number of atoms is confined by some external potential such that Bose-Einstein condensation may also occur in low dimensions [134].

5.4 Measuring the atomic field

The sum over $\mathbf{G}$ goes over all reciprocal lattice vectors $\mathbf{G} = \frac{2\pi}{d_0}(n_y \hat{\mathbf{e}}_y + n_z \hat{\mathbf{e}}_z)$ with n_j being some integer and $\omega_{q+k_x+\mathbf{G}} = \frac{\hbar}{2m}\left((k_x+q)^2 + \mathbf{G}^2\right)$. For simplicity we assume that $\mathbf{q}$ is parallel to the distance vector $\mathbf{d}$ defined in Eq. (5.1), such that the outcoupled atoms get a recoil momentum perpendicular to the atomic planes in the positive x-direction as indicated in Fig. (5.9). In this case $F_{L\to R}(t)$ is the only contribution to the interference flux $F_I(t)$. Using Eq. (5.86) in Eq. (5.26b) we find

$$\frac{F_{L\to R}(t)}{L^2} = \frac{4\pi^2 \xi w_0 \langle b \rangle \sqrt{n_L} \Gamma}{d_0^2} \quad (5.89)$$

$$\times \int dk_x \cos[\varphi_{LR} + k_x d - \delta\mu t + (\Omega - \omega_{q+k_x})t/2]\delta^{t/2}(\Omega - \omega_{q+k_x})e^{-k_x^2\xi^2}.$$

We note that the interference current is directly proportional to the superfluid order parameter of the optical lattice $\langle b \rangle$ and can therefore be used to monitor the quantum phase transition between Mott-insulator and superfluid phase. Moreover the interference contribution $F_{L\to R}(t)$ is accessible in an experiment by taking the time average of the measured photon signal as shown in Eq. (5.73). We now evaluate $F_{L\to R}(t)$ by numerically integrating Eq. (5.89). The superfluid order parameter $\langle b \rangle$ is thereby determined numerically within the mean-field treatment of the Bose-Hubbard model introduced in Sec. (2.3.1).

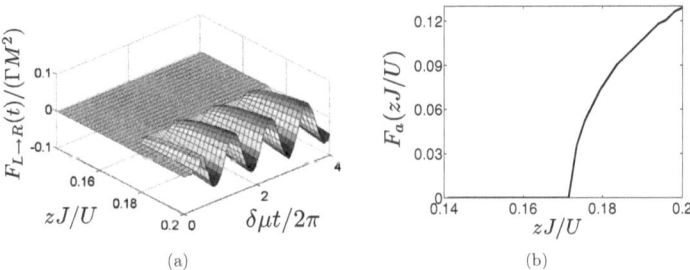

(a) (b)

Figure 5.10: a) Interference contribution of the photon flux $F_{L\to R}(t)$ in units of ΓM^2 as a function of time in units of $\frac{2\pi}{\delta\mu}$ and hopping zJ in units of U. b) Amplitude F_a as a function of hopping zJ in units of U. The parameters are $V_0 = 10E_R$, $\xi_x = 0.13 d_0$, $\delta\mu = 0.2\omega_R$, $d = 20 d_0$, $q = 2\pi/d_0$, $\Omega - \omega_q = \omega_R$ and $\varphi_{LR} = 0$. We use for the chemical potential of the optical lattice $\mu_R = \sqrt{2} - 1$ and $n_L = \frac{\langle b \rangle^2 |_{zJ/U=0.2}}{d_0^2} \approx 0.2/d_0^2$.

Figure (5.10a) displays the interference contribution $F_{L\to R}(t)$ as a function of time and hopping strength compared to the onsite interaction zJ/U. We assume constant lattice depth V_0 and the parameter that is changed in Fig. (5.10) is the onsite interaction strength U. For small values of zJ/U no interference signal is present since the atoms are in the Mott-insulator state. At a critical value zJ/U_c the right system undergoes the quantum phase transition to the superfluid state and the interference photon flux $F_{L\to R}(t)$ starts to oscillate in time with finite amplitude. We chose $\mu_R = \sqrt{2} - 1$ such that the value at which $F_{L\to R}(t)$ becomes nonzero corresponds to the tip of the first Mott lobe in Fig. (2.6a). In order to get a better understanding of the behaviour of $F_{L\to R}(t)$ as a function of zJ/U we define the amplitude

$$F_a(zJ/U) = \max(F_{L\to R}(t)) - \min(F_{L\to R}(t)), \quad (5.90)$$

where $F_{L\to R}(t)$ is taken as a function of time t only keeping all other parameters fixed and the minimum and maximum is taken over a time interval large compared to the inverse oscillation frequency of $F_{L\to R}(t)$, similar to Eq. (5.81).

In Fig. (5.10b) we plot the amplitude $F_a(zJ/U)$ as a function of the ratio zJ/U. From Eq. (5.89) we note that is directly proportional to the order parameter of the optical lattice $F_a(zJ/U) \propto \langle b \rangle$. For small values of zJ/U corresponding to the Mott-insulator phase for the atoms in the optical lattice $F_a(zJ/U) = 0$. At a critical value zJ/U_c the right system enters into the superfluid phase and the amplitude of $F_{L\to R}(t)$ becomes nonzero $F_a(zJ/U) > 0$. Hence measuring the scattered light from the atomic ensembles allows one to determine the quantum state of the ultracold atoms in the optical lattice. If the right system is in the Mott-insulator phase the measured light intensity is constant in time. In the superfluid phase the photon flux oscillates in time, where the oscillation frequency is determined by the difference in chemical potential between the left and right system and the amplitude of the oscillations is directly proportional to the superfluid order parameter. Hence the measurement of the quantum phase transition from Mott-insulator to superfluid state should be experimentally feasible, since the mere appearance of oscillations in the light signal indicates wether the right system is in the superfluid phase or not.

5.5 Discussion

In this chapter we have considered a setup where ultracold atoms confined into two spatially separated regions. They were outcoupled of their traps by two laser beams driving a Raman transition to an untrapped electronic state of the atoms. We thereby treated the light quantum mechanically and showed that the scattered light intensity oscillates in time if both systems have nonzero mean value of the atomic field operator. It was shown that the oscillation of the light intensity is due to a Josephson type coupling between the two macroscopic wave functions of the trapped atoms mediated by the outcoupled atoms. Applying our scheme to the specific experimental setup in [47] we find agreement between theory and the experimental results.

In Sec. (5.4) we argued that the presented scheme is analogue to a homodyne detection in optics, where the field to measure is mixed at a beam-splitter with a local

5.5 Discussion

oscillator. In the present case the Raman lasers act like a beam-splitter for the atomic fields and a Bose-Einstein condensate at very low temperatures $T \approx 0$ with known number of atoms N, such that $\langle \psi \rangle \approx \sqrt{N}$, may serve as a reference system and play the role of the local oscillator. It was shown that it is possible to measure the mean value of the atomic field operator of any other system formed by the same species of atoms by means of photon counting. We then considered two specific setups and showed how one may be able to measure the temperature of a Bose-Einstein condensate or monitor the phase transition from superfluid to Mott-insulator of ultracold atoms in an optical lattice by detecting the scattered light from the system.

It is important to note that the measurement of the scattered light intensity is an in situ measurement and only partly destructive since only a small part of the atoms have to be kicked out of the systems [47].

Chapter 6
Summary and Outlook

In the present thesis we have studied ultracold bosonic ensembles interacting dispersively with the electromagnetic field, and analyzed what kind of information about the many-body state of the atoms may be obtained by measuring the scattered photons. We first introduced the basic theoretical tools needed for our investigations. We then studied the photonic properties of a one dimensional string of point-like dipoles at fixed positions in space in a periodic configuration, interacting with the quantized electromagnetic field. It was shown that the transmission spectrum of a weak probe field could be used to obtain information about the spatial array of the atoms. Taking into account the external degrees of freedom of the dipoles we were led to consider a gas of ultracold atoms in a one dimensional optical lattice illuminated by a weak probe beam. It was shown that frequency resolved measurement of the scattered light at different angles could reveal the spectrum of the atomic system. The information obtained in such a Bragg scattering experiment is the structure form factor of the atomic system which is essentially the Fourier transform of the density-density correlation function of the atoms [44]. We then considered ultracold atoms confined in two spatially separated regions which were driven by two lasers and outcoupled of their traps due to a Raman transition to an untrapped electronic state. We treated the light quantum mechanically and demonstrated that, using the atoms in one trap as reference system with known order parameter $\langle \psi(\mathbf{r}, t) \rangle$, the order parameter of the atomic gas in the other trap may be determined by measuring the scattered light. We pointed out the analogy to optical homodyne detection and showed how this setup may allow for measuring the temperature of a Bose-Einstein condensate or monitor the superfluid to Mott-insulator phase transition of ultracold atoms in an optical lattice by means of photon counting. A detailed discussion of the main results was given at the end of each chapter.

We now give a general outlook to the work presented in this thesis. The possibility to reveal the quantum properties of matter in the scattered light is indeed attractive. Ideally, one would identify an interferometric setup which allows one to access several orders of correlation functions, as one usually does for light. In this respect, previous and recent works studied setups, based on detection of certain correlation functions of matter by means of elastically scattered photons. Due to the dispersive interaction such setups may be non-destructive for the quantum state of the atoms [35, 40, 41, 42], thus realizing quantum non-demolition measurements. Quantum non-demolition (QND) measurements have been introduced originally in order to overcome the standard quantum limit in a measurement process [135]. The measurement of the quadrature amplitude of a travelling electromagnetic wave by Levenson *et al.* in 1986 [136] constitutes the first experimental realization of a QND-measurement. Another example for the experimental realization of a QND-measurement is the detection of the collective spin states of an atomic ensemble using its dispersive interaction with light [137]. Applying this scheme to ultracold gases, it was shown theoretically that one may determine their magnetic properties in a non-destructive way [40, 42]. Following these approaches, it is intriguing to ask whether it is possible to extend the scheme presented in Chapter 5 to implement QND-measurements for the nondiagonal elements of the atomic density matrix which might allow for a complete determination of the quantum state of the atomic system.

Placing an ultracold gas inside a cavity may allow one to determine the quantum state of the atoms interacting with the cavity field in a non-destructive way by measuring the photons leaking out of the cavity [35, 33]. Several setups which use photo detection in order to obtain information about the quantum state of the atomic gas have already been realized experimentally. By placing a Bose-Einstein condensate inside a single mode optical resonator, the vacuum Rabi splitting due to the collective coupling of the condensate atoms with the cavity field was observed by pumping the cavity and measuring the transmitted light intensity as a function of frequency [138, 139]. Pumping directly the atoms a phase transition to a self-organized supersolid atomic state was observed by detecting the emitted light from the cavity [140]. In the regime where thermal fluctuations of the atoms are negligible it has been demonstrated theoretically that this system exhibits novel quantum phases which may be revealed by photo detection [141]. It is interesting to ask whether these setups allow one to implement feedback mechanisms onto the atomic gas, which could be used to engineer exotic quantum states of matter and lead to a transfer of the quantum state of matter onto the light and vice versa.

Another context where studying ultracold atomic gases by measuring the scattered light may be relevant is the study of nonequilibrium dynamics of manybody systems, which constitutes a direction of growing interest in the field of ultracold atomic gases. The identification of a QND-measurement for the relevant parameters would permit the study of the atomic dynamics in real time.

Appendix

Appendix A

Derivation of Eq. (5.15)

We consider the general case in which the dynamics of the system is given by the Hamiltonian

$$H = H_0 + V, \qquad (A.1)$$

where H_0 describes the unperturbed dynamics and V is a perturbation. We now outline how one obtains $\frac{d}{dt}\langle O \rangle$ for some operator O by expanding the density operator up to second order in the interaction V. The expectation value of O is given by

$$\langle O \rangle = \text{tr}\{\rho_{H_0} O_{H_0}\}, \qquad (A.2)$$

where we express the density matrix ρ_{H_0} and the operator O_{H_0} in the interaction picture with respect to H_0. The respective time evolution is given by [142]

$$\frac{\partial}{\partial t}\rho_{H_0}(t) = \frac{i}{\hbar}[\rho_{H_0}(t), V_{H_0}(t)], \qquad (A.3a)$$

$$\frac{d}{dt}O_{H_0} = \frac{i}{\hbar}[H_0, O_{H_0}]. \qquad (A.3b)$$

The formal solution to Eq. (A.3a) is given by

$$\begin{aligned}\rho_{H_0}(t) &= \rho_{H_0}(0) + \frac{i}{\hbar}\int_0^t dt'\,[\rho_{H_0}(t'), V_{H_0}(t')] \\ &\approx \rho_{eq} + \frac{i}{\hbar}\int_0^t dt'\,[\rho_{eq}, V_{H_0}(t')], \end{aligned} \qquad (A.4)$$

where ρ_{eq} is the density matrix at time 0. We focus on conserved quantities of the unperturbed system, $[H_0, O_{H_0}] = 0$ such that the time evolution of O is solely due to the interaction V,

$$\frac{d}{dt}\langle O \rangle = \frac{i}{\hbar}\text{tr}\{\rho_{H_0}(t)[V_{H_0}(t), O_{H_0}(t)]\}. \qquad (A.5)$$

Derivation of Eq. (5.15)

Using Eq. (A.4) in Eq. (A.5) one finds for the time derivation of $\langle O \rangle$ up to second order in V

$$\frac{d}{dt}\langle O \rangle = \frac{i}{\hbar} tr\left\{\rho_{eq}[V_{H_0}(t), O_{H_0}]\right\} - \frac{1}{\hbar^2}\int_0^t dt'\, tr\left\{\rho_{eq}[V_{H_0}(t'),[V_{H_0}(t), O_{H_0}]]\right\}. \quad (A.6)$$

Setting $O = a_2^\dagger a_2$ and using the perturbation V given in Eq. (5.7) one obtains from Eq. (A.6) the photon flux $F(t)$, Eq. (5.15).

Appendix B

Thermodynamics of trapped Bose-Einstein condensates

Here we will review some properties of a harmonically trapped Bose-Einstein condensate and derive the excitation spectrum using the local density approximation.

The atomic dynamics in the grand canonical ensemble is governed by the Hamiltonian

$$K = H - \mu N = \int d\mathbf{r} \psi^\dagger(\mathbf{r}) \left(\frac{-\hbar^2 \nabla^2}{2m} + V(\mathbf{r}) - \mu \right) \psi(\mathbf{r}) + \frac{g}{2} \int d\mathbf{r} \psi^\dagger(\mathbf{r}) \psi^\dagger(\mathbf{r}) \psi(\mathbf{r}) \psi(\mathbf{r}), \quad (B.1)$$

where the Heisenberg equation of motion for the atomic field operator is given by

$$i\hbar \frac{\partial}{\partial t} \psi(\mathbf{r}, t) = \left(\frac{-\hbar^2 \nabla^2}{2m} + V(\mathbf{r}) - \mu \right) \psi(\mathbf{r}, t) + g \psi^\dagger(\mathbf{r}, t) \psi(\mathbf{r}, t)^2. \quad (B.2)$$

Separating out the condensate part of the wavefunction

$$\psi(\mathbf{r}, t) = \Phi(\mathbf{r}) + \delta\psi(\mathbf{r}, t), \quad (B.3)$$
$$\Phi(\mathbf{r}) = \langle \psi(\mathbf{r}, t) \rangle, \quad (B.4)$$

we find for the term arising from s-wave scattering of the atoms in Eq. (B.2)

$$\psi^\dagger \psi \psi = |\Phi|^2 \Phi + 2|\Phi|^2 \delta\psi + \Phi^2 \delta\psi^\dagger + \Phi^* \delta\psi \delta\psi + 2\Phi \delta\psi^\dagger \delta\psi + \delta\psi^\dagger \delta\psi \delta\psi. \quad (B.5)$$

The last term on the right hand side can be treated in a self consistent meanfield approximation [143]

$$\delta\psi^\dagger \delta\psi \delta\psi \approx 2 \left\langle \delta\psi^\dagger \delta\psi \right\rangle \delta\psi + \langle \delta\psi \delta\psi \rangle \delta\psi^\dagger = 2 n_T(\mathbf{r}) \delta\psi(\mathbf{r}) + m_T(\mathbf{r}) \delta\psi^\dagger, \quad (B.6)$$

where

$$n_T(\mathbf{r}) = \left\langle \delta\psi^\dagger \delta\psi \right\rangle, \quad (B.7a)$$
$$m_T(\mathbf{r}) = \langle \delta\psi \delta\psi \rangle, \quad (B.7b)$$

are the normalous and anormalous densities respectively. We now introduce for later convenience the densities

$$n_0(\mathbf{r}) = |\Phi(\mathbf{r})|^2, \tag{B.8a}$$
$$m_0(\mathbf{r}) = \Phi(\mathbf{r})^2, \tag{B.8b}$$
$$n(\mathbf{r}) = n_0(\mathbf{r}) + n_T(\mathbf{r}), \tag{B.8c}$$
$$m(\mathbf{r}) = m_0(\mathbf{r}) + m_T(\mathbf{r}). \tag{B.8d}$$

Using Eq. (B.6) in Eq. (B.5) one arrives at

$$\psi^\dagger \psi \psi = |\Phi|^2 \Phi + 2\left(|\Phi|^2 + n_T\right)\delta\psi + \left(\Phi^2 + m_T\right)\delta\psi^\dagger + \Phi^*\delta\psi\delta\psi + 2\Phi\delta\psi^\dagger\delta\psi. \tag{B.9}$$

Interting Eq. (B.9) in Eq. (B.2) and taking the expectation value one finds

$$0 = \left(\frac{-\hbar^2 \nabla^2}{2m} + V(\mathbf{r}) - \mu + g\left(n_0(\mathbf{r}) + 2n_T(\mathbf{r})\right)\right)\Phi(\mathbf{r}) + m_T(\mathbf{r})\Phi^*. \tag{B.10}$$

If we neglect $n_T(\mathbf{r})$ and $m_T(\mathbf{r})$ Eq. (B.10) reduces to the standard Gross-Pitaevskii equation in the Bogoliubov approximation [61]. The equation of motion for the excitations of the condensate $\delta\psi(\mathbf{r},t)$ can be obtained by subtracting Eq. (B.10) from Eq. (B.2) and is given by

$$i\hbar\frac{\partial}{\partial t}\delta\psi(\mathbf{r},t) = \mathcal{L}\delta\psi(\mathbf{r},t) + gm(\mathbf{r})\delta\psi^\dagger, \tag{B.11}$$

where the quadratic terms have been treated consistent within the meanfield approximation [143]

$$\delta\psi\delta\psi - \langle\delta\psi\delta\psi\rangle = 0, \tag{B.12}$$
$$\delta\psi^\dagger\delta\psi - \langle\delta\psi^\dagger\delta\psi\rangle = 0, \tag{B.13}$$

and the operator $\mathcal{L}$ reads

$$\mathcal{L} = \left(\frac{-\hbar^2 \nabla^2}{2m} + V(\mathbf{r}) - \mu + 2gn(\mathbf{r})\right). \tag{B.14}$$

A solution of Eq. (B.11) can be obtained by the following Bogoliubov Ansatz

$$\delta\psi(\mathbf{r},t) = \sum_j \left(u_j(\mathbf{r})\alpha_j e^{-i\omega_j t} - v_j^*(\mathbf{r})\alpha_j^\dagger e^{i\omega_j t}\right), \tag{B.15}$$

where the operators $\alpha, \alpha^\dagger$ are annihilation and creation operators of noninteracting quasiparticles with the standard Bose commutation relation $\left[\alpha_j, \alpha_k^\dagger\right] = \delta_{j,k}$, and the functions $u_j(\mathbf{r}), v_j(\mathbf{r})$ obey the normalization relation

$$\int d\mathbf{r}\left(u_i^*(\mathbf{r})u_j(\mathbf{r}) - v_i^*(\mathbf{r})v_j(\mathbf{r})\right) = \delta_{i,j}. \tag{B.16}$$

B.1 Non interacting case

The Bogoliubov Ansatz Eq. (B.15) yields a solution of Eq. (B.11), if the functions $u_j(\mathbf{r}), v_j(\mathbf{r})$ satisfy the Hatree-Fock-Bogoliubov (HFB) equations

$$\mathcal{L}(\mathbf{r})u_j(\mathbf{r}) - gm(\mathbf{r})v_j(\mathbf{r}) = \hbar\omega_j u_j(\mathbf{r}), \qquad \text{(B.17a)}$$
$$-\mathcal{L}(\mathbf{r})v_j(\mathbf{r}) + gm(\mathbf{r})^* u_j(\mathbf{r}) = \hbar\omega_j v_j(\mathbf{r}). \qquad \text{(B.17b)}$$

The normalous and anomalous densities defined in Eq. (B.7) are given by

$$n_T(\mathbf{r}) = \sum_j \left[\left(|u_j(\mathbf{r})|^2 + |v_j(\mathbf{r})|^2 \right) N_0(\omega_j) + |v_j(\mathbf{r})|^2 \right], \qquad \text{(B.18)}$$

$$m_T(\mathbf{r}) = -\sum_j u_j(\mathbf{r})v_j(\mathbf{r})^* \left(2N_0(\omega_j) + 1 \right), \qquad \text{(B.19)}$$

where $N_0(\omega)$ is the Bose distribution for the quasi particles

$$N_0(\omega) = \frac{1}{e^{\beta\hbar\omega} - 1} = \left\langle \alpha_j^\dagger \alpha_j \right\rangle. \qquad \text{(B.20)}$$

In the following we will discuss the case of a harmonic potential.

$$V(\mathbf{r}) = \frac{1}{2}m(\omega_x^2 x^2 + \omega_y^2 y^2 + \omega_z^2 z^2) \qquad \text{(B.21)}$$

B.1 Non interacting case

In the case of noninteracting particles $g = 0$ one has $v_j(\mathbf{r}) = 0$ and Eqs. (B.17) reduce to the Schrödinger equation of the 3 dimensional harmonic oscillator with frequencies $\omega_{\{x,y,z\}}$ and

$$\hbar\omega_j = \hbar\left(\omega_x(n_x + \frac{1}{2}) + \omega_y(n_y + \frac{1}{2}) + \omega_z(n_z + \frac{1}{2}) \right) - \mu. \qquad \text{(B.22)}$$

The number of atoms not in the condensate can be obtained from $N_{\text{ex}}(T) = \int d\mathbf{r}\, n_T(\mathbf{r}) = \sum_j \frac{1}{e^{\beta\hbar\omega_j} - 1}$, where the chemical potential was taken equal to the lowest energy state [122]. This leads to

$$\begin{aligned} N_{\text{ex}}(T) &= \sum_{n_x,n_y,n_z} \frac{1}{e^{\hbar\beta(n_x+n_y+n_z)} - 1} \\ &\approx \frac{1}{(\hbar\bar{\omega})^3} \int dx\,dy\,dz \frac{1}{e^{\beta(x+y+z)} - 1} \\ &= \frac{\zeta(3)}{(\hbar\bar{\omega}\beta)^3}, \end{aligned} \qquad \text{(B.23)}$$

with $\bar{\omega} = (\omega_x\omega_y\omega_z)^{1/3}$ and $\zeta(x)$ is the Riemann Zeta function. The critical temperature T_c at which Bose-Einstein condensation takes place can be calculated from the condition that $N_{\text{ex}}(T_c) = N$, which leads to [61]

$$kT_c = \hbar\bar{\omega}\left(\frac{N}{\zeta(3)}\right)^{1/3} \approx 0.94\hbar\bar{\omega}N^{1/3}. \quad (B.24)$$

For the condensate fraction $N_C(T) = N - N_{\text{ex}}(T)$ one finds

$$N_C(T) = N\left(1 - \frac{T^3}{T_c^3}\right). \quad (B.25)$$

The macroscopic wavefunction of the condensate $\Phi(\mathbf{r})$ is given by the ground state wavefunction of a harmonic oscillator with the normalization $\int d\mathbf{r}|\Phi(\mathbf{r})|^2 = N_C(T)$, such that

$$\Phi(\mathbf{r}) = \sqrt{N_C(T)}\left(\frac{m\bar{\omega}}{\pi\hbar}\right)^{3/4} \exp\left[-\frac{m}{2\hbar}(\omega_x x^2 + \omega_y y^2 + \omega_z z^2)\right]. \quad (B.26)$$

B.2 Interacting case

In the case of interacting particles $g \neq 0$ Eqs. (B.17) have to be solved numerically in general. However, they can be solved analytically using the local density approximation (LDA), with the following Ansatz for the Bogoliubov amplitudes [144]

$$u_j(\mathbf{r}) = u_{\mathbf{k}}(\mathbf{r})e^{i\mathbf{k}\cdot\mathbf{r}}, \quad (B.27)$$
$$v_j(\mathbf{r}) = v_{\mathbf{k}}(\mathbf{r})e^{i\mathbf{k}\cdot\mathbf{r}}. \quad (B.28)$$

The different excitations are now labeled by their momentum $\hbar\mathbf{k}$ and $\sum_j \to \frac{1}{(2\pi)^3}\int d\mathbf{k}$. In the LDA one replaces the solutions of the trapped system with the solutions in the free space case adjusted to the local density given by the trap. In the free space case $u_{\mathbf{k}}(\mathbf{r}), v_{\mathbf{k}}(\mathbf{r})$ are independent on position and hence we assume them to be slowly varying functions of $\mathbf{r}$. Neglecting derivatives of $u_{\mathbf{k}}(\mathbf{r})$ and $v_{\mathbf{k}}(\mathbf{r})$, Eqs. (B.17) read

$$\left(\frac{\hbar^2\mathbf{k}^2}{2m} + V(\mathbf{r}) - \mu + 2gn(\mathbf{r})\right)u_{\mathbf{k}}(\mathbf{r}) - gm(\mathbf{r})v_{\mathbf{k}}(\mathbf{r}) = E_{\mathbf{k}}(\mathbf{r})u_{\mathbf{k}}(\mathbf{r}), \quad (B.29)$$
$$-\left(\frac{\hbar^2\mathbf{k}^2}{2m} + V(\mathbf{r}) - \mu + 2gn(\mathbf{r})\right)v_{\mathbf{k}}(\mathbf{r}) + gm(\mathbf{r})^*u_{\mathbf{k}}(\mathbf{r}) = E_{\mathbf{k}}(\mathbf{r})v_{\mathbf{k}}(\mathbf{r}). \quad (B.30)$$

B.2 Interacting case

The normalization $|u_{\mathbf{k}}(\mathbf{r})|^2 - |v_{\mathbf{k}}(\mathbf{r})|^2 = 1$ leads to [144]

$$E_{\mathbf{k}}(\mathbf{r}) = \sqrt{\left(\frac{\hbar \mathbf{k}^2}{2m} + V(\mathbf{r}) - \mu + 2gn(\mathbf{r})\right)^2 - g|m(\mathbf{r})|^2}, \quad \text{(B.31a)}$$

$$|u_{\mathbf{k}}(\mathbf{r})|^2 = \frac{\frac{\hbar^2 \mathbf{k}^2}{2m} + V(\mathbf{r}) - \mu + 2gn(\mathbf{r})}{2E_{\mathbf{k}}(\mathbf{r})} + \frac{1}{2}, \quad \text{(B.31b)}$$

$$|v_{\mathbf{k}}(\mathbf{r})|^2 = \frac{\frac{\hbar^2 \mathbf{k}^2}{2m} + V(\mathbf{r}) - \mu + 2gn(\mathbf{r})}{2E_{\mathbf{k}}(\mathbf{r})} - \frac{1}{2}. \quad \text{(B.31c)}$$

In the case $N_C a_s/\bar{a} \gg 1$, where a_s is the s-wave scattering length and $\bar{a} = \sqrt{\frac{\hbar}{m\bar{\omega}}}$, the interaction energy will be dominant over the kinetic energy. Neglecting the kinetic energy part in Eq. (B.10) is called the Thomas-Fermi approximation and one obtains [61]

$$\left(V(\mathbf{r}) - \mu + g(n_0(\mathbf{r}) + 2n_T(\mathbf{r}))\right)\Phi(\mathbf{r}) + m_T(\mathbf{r})\Phi^* = 0. \quad \text{(B.32)}$$

Neglecting $n_T(\mathbf{r})$ and $m_T(\mathbf{r})$, which is equivalent to the Bogoliubov approximation, the Thomas-Fermi profile for the condensate reads

$$\Phi(\mathbf{r}) = \begin{cases} \sqrt{\frac{1}{g}(\mu - V(\mathbf{r}))} & \text{for } \mathbf{r} \in V_{\text{cond}}, \\ 0 & \text{for } \mathbf{r} \notin V_{\text{cond}}. \end{cases} \quad \text{(B.33)}$$

One can distinguish two regions of space depending on wether the coordinate $\mathbf{r}$ lies inside or outside the volume V_{cond}, giving the region to which the condensate is confined. The volume V_{cond} is determined by the condition $\mu - V(\mathbf{r}) \geq 0$. The chemical potential in Eq. (B.33) is determined by the condition $\int_{V_{\text{cond}}} d\mathbf{r} |\Phi(\mathbf{r})|^2 = N_C$ which yields for the harmonic potential Eq. (B.21) [61]

$$\mu = \frac{15^{2/5}}{2}\left(\frac{N_U a_s}{\bar{a}}\right)^{2/5} \hbar \bar{\omega}$$

$$\approx \eta k T_c (1 - t_r^3)^{2/5}, \quad \text{(B.34)}$$

where $t_r = T/T_c$ is the reduced temperature, we used Eq. (B.25) for the noncondensed atoms and introduced the parameter

$$\eta = \frac{\mu(T=0)}{kT_c}$$

$$\approx 1.57 \left(N^{1/6} \frac{a_s}{\bar{a}}\right)^{2/5}. \quad \text{(B.35)}$$

In Eq(B.31) we set [122]

$$n(\mathbf{r}) = m(\mathbf{r}) = 0 \quad \text{for } \mathbf{r} \notin V_{\text{cond}},$$
$$g\, n(\mathbf{r}) = \mu - V(\mathbf{r}) \quad \text{for } \mathbf{r} \in V_{\text{cond}},$$
$$n(\mathbf{r})^2 = |m(\mathbf{r})|^2 \quad \text{for } \mathbf{r} \in V_{\text{cond}}.$$

which leads to

$$E_{\mathbf{k}}(\mathbf{r}) = \begin{cases} \sqrt{\varepsilon_{\mathbf{k}}^2 + 2\varepsilon_{\mathbf{k}}(\mu - V(\mathbf{r}))} & \text{for } \mathbf{r} \in V_{\text{cond}}, \\ \varepsilon_{\mathbf{k}} + V(\mathbf{r}) - \mu & \text{for } \mathbf{r} \notin V_{\text{cond}}, \end{cases} \quad \text{(B.36a)}$$

$$|u_{\mathbf{k}}(\mathbf{r})|^2 = \begin{cases} \frac{\varepsilon_{\mathbf{k}} + \mu - V(\mathbf{r})}{2E_{\mathbf{k}}(\mathbf{r})} + \frac{1}{2} & \text{for } \mathbf{r} \in V_{\text{cond}}, \\ 1 & \text{for } \mathbf{r} \notin V_{\text{cond}}, \end{cases} \quad \text{(B.36b)}$$

$$|v_{\mathbf{k}}(\mathbf{r})|^2 = \begin{cases} \frac{\varepsilon_{\mathbf{k}} + \mu - V(\mathbf{r})}{2E_{\mathbf{k}}(\mathbf{r})} - \frac{1}{2} & \text{for } \mathbf{r} \in V_{\text{cond}}, \\ 0 & \text{for } \mathbf{r} \notin V_{\text{cond}}, \end{cases} \quad \text{(B.36c)}$$

where for the harmonic potential Eq. (B.21) the chemical potential μ is given by Eq. (B.34). The number of noncondensed atoms is given by

$$\begin{aligned} N_{\text{ex}} &= \int d\mathbf{r}\, n_T(\mathbf{r}) \\ &= \int d\mathbf{r} \int \frac{d\mathbf{k}}{(2\pi)^3} \left[\left(|u_{\mathbf{k}}(\mathbf{r})|^2 + |v_{\mathbf{k}}(\mathbf{r})|^2\right) N_0(E_{\mathbf{k}}(\mathbf{r})/\hbar) + |v_{\mathbf{k}}(\mathbf{r})|^2 \right], \end{aligned} \quad \text{(B.37)}$$

which has to be evaluated numerically in a self consistent manner. Eq. (B.37) may be evaluated to lowest order in a_s by noting that most noncondensed atoms will be outside the condensate [122]. Using expressions Eqs. (B.36) for $\mathbf{r} \notin V_{\text{cond}}$ one finds

$$\begin{aligned} N_{\text{ex}} &\approx \int d\mathbf{r} \int \frac{d\mathbf{k}}{(2\pi)^3} \frac{1}{\exp\left[\beta(\varepsilon_{\mathbf{k}} + V(\mathbf{r}) - \mu)\right] - 1} \\ &= \frac{1}{(\hbar\bar{\omega})^3} \sum_{k=1}^{\infty} \frac{e^{\beta\mu k}}{k^3} \\ &\approx \frac{1}{(\hbar\bar{\omega})^3} \left(\zeta(3) + \beta\mu\zeta(2)\right), \end{aligned} \quad \text{(B.38)}$$

where we expanded the exponential up to lowest order in μ. We note that the first term agrees with the noninteracting result Eq. (B.23) and the second term is the first correction due to the interactions. Using Eq. (B.34) for the chemical potential we then find for the condensed fraction

$$N_C(T) = N \left(1 - t_r^3 - \frac{\zeta(2)}{\zeta(3)} t_r^2 \eta (1 - t_r^3)^{2/5}\right). \quad \text{(B.39)}$$

Although the higher order terms in Eq. (B.38) diverge Eq. (B.39) gives good results for the noncondensed fraction [61]. The critical temperature T_c^i for an interacting Bose-Einstein condensate can now be determined by setting $N_C(T_c^i) = 0$ for a given parameter η.

Appendix C

Derivation of Eq. (5.56)

In the following we will derive approximate expressions of the Raman scattering rate for the eperimental setup of [47] which lead to Eq. (5.56).

Using Eq. (5.52) for the condensate wave functions in Eq. (5.49) we find

$$
\begin{aligned}
F_L(t) &\approx \frac{\mu \Gamma}{g} \int_0^{\min(t, 2r_x/v_q)} dt' \cos\left[(\omega_{\mathbf{q}} - \Omega) t'\right] \\
&\quad \times \int d\mathbf{r} \left(1 - \frac{x^2}{r_x^2} - \frac{y^2}{r_y^2} - \frac{z^2}{r_z^2}\right)^{1/2} \left(1 - \frac{(x - v_q t')^2}{r_x^2} - \frac{y^2}{r_y^2} - \frac{z^2}{r_z^2}\right)^{1/2} \\
&= \frac{2\pi \Gamma \mu r_x \bar{R}^3}{g v_q} \int_0^{z_<(0)} dz \cos\left[(\omega_{\mathbf{q}} - \Omega) \frac{z r_x}{v_q}\right] G(z) ,
\end{aligned}
\qquad (C.1)
$$

with

$$
z_<(x) = \min\left(2, \frac{tv_q - x}{r_x}\right) . \qquad (C.2)
$$

The length $\bar{R} = (r_x r_y r_z)^{1/3}$ determines the typical size of the condensates and can be written as [122]

$$
\bar{R} = 15^{1/5} \left(\frac{N a_s}{\bar{a}}\right)^{1/5} \bar{a} . \qquad (C.3)
$$

The function $G(z)$ in Eq. (C.1) is given by

$$
G(z) = \int dx \int_{r>0} r \, dr \, (1 - x^2 - r^2)^{1/2} \left(1 - (x - z)^2 - r^2\right)^{1/2} , \qquad (C.4)
$$

where the integration boundaries are chosen such that the arguments in the square roots are always positive for $r > 0$. The boundaries of the integration, where the two integrands vanish, are circles of radius 1 around $(0,0)$ and $(z,0)$ in the $x - r$ plane as

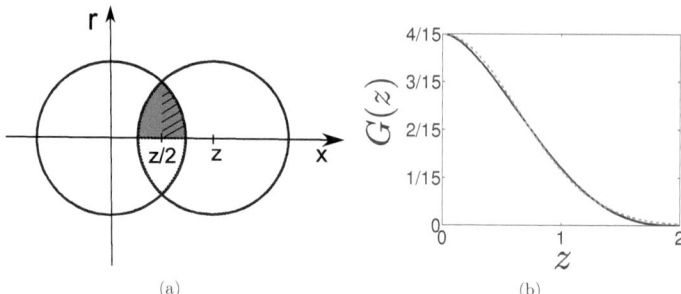

Figure C.1: a) The two circles indicate the regions where the integrands in Eq. (C.4) are different from zero, such that the whole integral goes over the blue filled region. Due to the symmetry of the wave functions we can restrict to the blue shaded area. b) Comparison of the integral $G(z)$ as given in Eq. (C.5) (black solid line) with the Gaussian fit Eq. (C.6) (red dashed line).

shown in Fig. (C.1a). Hence $G(z)$ can be written as

$$G(z) = 2 \int_{z/2}^{1} dx \int_{0}^{\sqrt{1-x^2}} r\, dr \left(1 - x^2 - r^2\right)^{1/2} \left(1 - (x-z)^2 - r^2\right)^{1/2}, \qquad (\text{C.5})$$

where the symmetry around $x = z/2$ and the integration region is indicated in Fig. (C.1a). A numerical evaluation of $G(z)$ is shown in Fig. (C.1b) compared to the Gaussian approximation

$$G(z) \approx \frac{4}{15} e^{-1.25 z^2}. \qquad (\text{C.6})$$

Using Eq. (C.6) in Eq. (C.1) and the corresponding expressions for $F_R(t)$ and the interference contributions one get

$$F_L(t) \approx \frac{8F_0}{15\pi} \int_0^{z_<(0)} dz \cos\left[(\omega_q - \Omega)zt_0\right] e^{-1.25z^2}, \qquad (C.7a)$$

$$F_R(t) \approx \frac{8F_0}{15\pi} \int_0^{z_<(0)} dz \cos\left[(\omega_q - \Omega + \delta\mu)zt_0\right] e^{-1.25z^2}, \qquad (C.7b)$$

$$F_{L\to R}(t) \approx \frac{8F_0}{15\pi} \mathrm{Re}\, e^{i(\delta\mu t - \varphi_{LR} + (\omega_q - \Omega)\frac{d}{v_q})}$$
$$\times \int_{-2}^{z_<(d)} dz e^{i(\omega_q - \Omega)zt_0} e^{-1.25z^2} \Theta(q)\Theta\left(t - \frac{d - 2r_x}{v_q}\right), \qquad (C.7c)$$

$$F_{R\to L}(t) \approx \frac{8F_0}{15\pi} \mathrm{Re}\, e^{i(\delta\mu t - \varphi_{LR} - (\omega_q - \Omega + \delta\mu)\frac{d}{v_q})}$$
$$\times \int_{-2}^{z_<(d)} dz e^{i(\omega_q - \Omega - \delta\mu)zt_0} e^{-1.25z^2} \Theta(-q)\Theta\left(t - \frac{d - 2r_x}{v_q}\right), \qquad (C.7d)$$

with $t_0 = \frac{r_x}{v_q}$ and

$$F_0 = \frac{15\pi \Gamma t_0 N_C}{8}. \qquad (C.8)$$

For times $t > \frac{d+2r_x}{v_q}$, such that $z_<(0) = z_<(d) = 2$ we find from Eq. (C.7a)

$$F_L(t) \approx \Gamma t_0 N_C \int_0^2 dz \cos\left[(\omega_\mathbf{q} - \Omega)\frac{zr_x}{v_q}\right] e^{-1.25z^2}$$
$$= \frac{\pi^{1/2} \Gamma t_0 N_C}{2\sqrt{5}} e^{-\frac{1}{5}[(\omega_\mathbf{q}-\Omega)t_0]^2} \left[\mathrm{Erf}\left(\frac{5 - i(\omega_\mathbf{q}-\Omega)t_0}{\sqrt{5}}\right) + \mathrm{Erf}\left(\frac{5 + i(\omega_\mathbf{q}-\Omega)t_0}{\sqrt{5}}\right)\right]$$
$$\approx \pi \Gamma N_C \sqrt{\frac{t_0^2}{5\pi}} e^{-\frac{1}{5}[(\omega_\mathbf{q}-\Omega)t_0]^2}, \qquad (C.9)$$

where $\mathrm{Erf}(x)$ is the Error-function [145]. Neglecting the imaginary part in the Error-functions is a good approximation for $|(\omega_\mathbf{q} - \Omega)t_0| \geq 5$, as can be checked by numerical evaluation. In this case the exponential is negligible compared to the resonant case $\omega_\mathbf{q} = \Omega$. Thus we see that the oscillating tails of the Error-functions given by their imaginary parts will only play a role for parameters where the outcoupling efficiency vanishes. For the other contributions to the photon flux we find

$$F_L(t) \approx \pi \Gamma N_C K(\omega_{\mathbf{q}} - \Omega), \tag{C.10a}$$

$$F_R(t) \approx \pi \Gamma N_C K(\omega_{\mathbf{q}} - \Omega + \delta\mu), \tag{C.10b}$$

$$F_{L \to R}(t) \approx 2\pi \Gamma N_C K(\omega_{\mathbf{q}} - \Omega)$$
$$\times \cos\left[\delta\mu t - \varphi_{LR} + (\omega_{\mathbf{q}} - \Omega)\frac{d}{v_q}\right] F_\Theta(q,t), \tag{C.10c}$$

$$F_{R \to L}(t) \approx 2\pi \Gamma N_C K(\omega_{\mathbf{q}} - \Omega + \delta\mu)$$
$$\times \cos\left[\delta\mu t - \varphi_{LR} - (\omega_{\mathbf{q}} - \Omega + \delta\mu)\frac{d}{v_q}\right] F_\Theta(-q,t), \tag{C.10d}$$

with

$$K(x) = \sqrt{\frac{t_0^2}{5\pi}} e^{-\frac{t_0^2}{5}x^2}, \tag{C.11}$$

$$F_\Theta(q,t) = \Theta(q)\Theta\left(t - \frac{d - 2r_x}{v_q}\right). \tag{C.12}$$

From Eqs. (C.10) one then obtains Eq. (5.56) for $q > 0$.

Bibliography

[1] M. H. Anderson, J. R. Ensher, M. R. Matthews, C. E. Wieman, and E. A. Cornell. Observation of bose-einstein condensation in a dilute atomic vapor. *Science*, 269:198, 1995.

[2] K. B. Davis, M. O. Mewes, M. R. Andrews, N. J. van Druten, D. S. Durfee, D. M. Kurn, and W. Ketterle. Bose-einstein condensation in a gas of sodium atoms. *Phys. Rev. Lett.*, 75(22):3969–3973, 1995.

[3] C. C. Bradley, C. A. Sackett, J. J. Tollett, and R. G. Hulet. Evidence of bose-einstein condensation in an atomic gas with attractive interactions. *Phys. Rev. Lett.*, 75(9):1687–1690, 1995.

[4] M. R. Andrews, C. G. Townsend, H. J. Miesner, D. S. Durfee, D. M. Kurn, and W. Ketterle. Observation of interference between two bose condensates. *Science*, 275:637, 1997.

[5] I. Bloch, T. W. Hänsch, and T. Esslinger. Measurement of the spatial coherence of a trapped bose gas at the phase transition. *Nature*, 403:166, 2000.

[6] M. R. Matthews, B. P. Anderson, P. C. Haljan, D. S. Hall, C. E. Wieman, and E. A. Cornell. Vortices in a bose-einstein condensate. *Phys. Rev. Lett.*, 83(13):2498–2501, 1999.

[7] K. W. Madison, F. Chevy, W. Wohlleben, and J. Dalibard. Vortex formation in a stirred bose-einstein condensate. *Phys. Rev. Lett.*, 84(5):806–809, 2000.

[8] J. R. Abo-Shaeer, C. Raman, J. M. Vogels, and W. Ketterle. Observation of vortex lattices in bose-einstein condensates. *Science*, 292:476, 2001.

[9] G. Magayar and L. Mandel. Interference fringes produced by superposition of two independent maser light beams. *Nature*, 198:255, 1963.

[10] R. L. Pfleegor and L. Mandel. Interference of independent photon beams. *Phys. Rev.*, 159(5):1084, 1967.

[11] D. Rozas, Z. S. Sacks, and G. A. Swartzlander. Experimental observation of fluidlike motion of optical vortices. *Phys. Rev. Lett.*, 79(18):3399, 1997.

[12] D. Jaksch, C. Bruder, J. I. Cirac, C. W. Gardiner, and P. Zoller. Cold bosonic atoms in optical lattices. *Phys. Rev. Lett.*, 81:3108, 1998.

[13] M. Greiner, O. Mandel, T. Esslinger, T. W. Hänsch, and I. Bloch. Quantum phase transition from a superfluid to a mott insulator in an optical lattice. *Nature*, 39:415, 2002.

[14] I. Bloch, J. Dalibard, and W. Zwerger. Many-body physics with ultracold gases. *Rev. Mod. Phys.*, 80:885, 2008.

[15] M. Lewenstein, A. Sanpera, V. Ahufinger, B. Damski, A. Sen De, and U. Sen. Ultracold atomic gases in optical lattices: mimicking condensed matter physics and beyond. *Adv. Phys.*, 56:243, 2007.

[16] J. Stenger, S. Inouye, A. P. Chikkatur, D. M. Stamper-Kurn, D. E. Pritchard, and W. Ketterle. Bragg spectroscopy of a bose-einstein condensate. *Phys. Rev. Lett.*, 82:4569, 1999.

[17] D. M. Stamper-Kurn, A. P. Chikkatur, A. Görlitz, S. Inouye, S. Gupta, D. E. Pritchard, and W. Ketterle. Excitation of phonons in a bose-einstein condensate by light scattering. *Phys. Rev. Lett.*, 83:2876, 1999.

[18] R. Ozeri, N. Katz, J. Steinhauer, and N. Davidson. Colloquium: Bulk bogoliubov excitations in a bose-einstein condensate. *Rev. Mod. Phys.*, 77:187, 2005.

[19] D. Clément, N. Fabbri, L. Fallani, C. Fort, and M. Inguscio. Exploring correlated 1d bose gases from the superfluid to the mott-insulator state by inelastic light scattering. *Phys. Rev. Lett.*, 102:155301, 2009.

[20] A. Brunello, F. Dalfovo, L. Pitaevskii, S. Stringari, and F. Zambelli. Momentum transferred to a trapped bose-einstein condensate by stimulated light scattering. *Phys. Rev. A*, 64:063614, 2001.

[21] A. M. Rey, P. B. Blakie, G. Pupillo, C. J. Williams, and C. W. Clark. Bragg spectroscopy of ultracold atoms loaded in an optical lattice. *Phys. Rev. A*, 72:023407, 2005.

[22] C. Menotti, M. Krämer, L. Pitaevskii, and S. Stringari. Dynamic structure factor of a bose-einstein condensate in a one-dimensional optical lattice. *Phys. Rev. A*, 67:053609, 2003.

[23] S. Ritter, A. Öttl, T. Donner, T. Bourdel, M. Köhl, and T. Esslinger. Observing the formation of long-range order during bose-einstein condensation. *Phys. Rev. Lett.*, 98(9):090402, 2007.

[24] N. R. Cooper and Z. Hadzibabic. Measuring the superfluid fraction of an ultracold atomic gas. *Phys. Rev. Lett.*, 104(3):030401, 2010.

[25] E. Altman, E. Demler, and M. D. Lukin. Probing many-body states of ultracold atoms via noise correlations. *Phys. Rev. A*, 70(1):013603, 2004.

[26] A. Polkovnikov, E. Altman, and E. Demler. Interference between independent fluctuating condensates. *Proc. Natl. Acad. Sci. USA*, 103:6125, 2006.

[27] S. Hofferberth, I. Lesanovsky, T. Schumm, A. Imambekov, V. Gritsev, E. Demler, and J. Schmiedmayer. Probing quantum and thermal noise in an interacting many-body system. *Nature Physics*, 4:489, 2008.

[28] S. Rath and W. Zwerger. Full counting statistics of the interference contrast from independent bose-einstein condensates. *arXiv:1009.4844 (cond-mat.quant-gas)*.

[29] S. Fölling, F. Gerbier, A. Widera, O. Mandel, T. Gericke, and I. Bloch. Spatial quantum noise interferometry in expanding ultracold atom clouds. *Nature*, 434:481, 2005.

[30] M. Greiner, C. A. Regal, J. T. Stewart, and D. S. Jin. Probing pair-correlated fermionic atoms through correlations in atom shot noise. *Phys. Rev. Lett.*, 94(11):110401, 2005.

[31] Z. Hadzibabic, P. Krüger, M. Cheneau, B. Battelier, and J. Dalibard. Berezinskii-kosterlitz-thouless crossover in a trapped atomic gas. *Nature*, 441:1118, 2006.

[32] I. B. Mekhov, C. Maschler, and H. Ritsch. Cavity-enhanced light scattering in optical lattices to probe atomic quantum statistics. *Phys. Rev. Lett.*, 98:100402, 2007.

[33] P. Meystre and M. Sargent III. *Elements of Quantum Optics*. Springer Verlag, 1991.

[34] P. Cañizares, T. Görler, J. P. Paz, G. Morigi, and W. P. Schleich. Signatures of non-locality in the first-order coherence of the scattered light. *Laser Physics*, 17:903, 2007.

[35] I. B. Mekhov, C. Maschler, and H. Ritsch. Probing quantum phases of ultracold atoms in optical lattices by transmission spectra in cavity qed. *Nature Phys.*, 3:319, 2007.

[36] J. Larson, B. Damski, G. Morigi, and M. Lewenstein. Mott-insulator states of ultracold atoms in optical resonators. *Phys. Rev. Lett.*, 100:050401, 2008.

[37] J. Ruostekoski, C. J. Foot, and A. B. Deb. Light scattering for thermometry of fermionic atoms in an optical lattice. *Phys. Rev. Lett.*, 103:170404, 2009.

[38] J. M. Higbie, L. E. Sadler, S. Inouye, A. P. Chikkatur, S. R. Leslie, K. L. Moore, V. Savalli, and D. M. Stamper-Kurn. Direct nondestructive imaging of magnetization in a spin-1 bose-einstein gas. *Phys. Rev. Lett.*, 95(5):050401, 2005.

[39] L. E. Sadler, J. M. Higbie, S. R. Leslie, M. Vengalattore, and D. M. Stamper-Kurn. Spontaneous symmetry breaking in a quenched ferromagnetic spinor bose-einstein condensate. *Nature*, 443:312, 2006.

[40] K. Ekert, O. Romero-Isart, M. Rodriguez, M. Lewenstein, E. Polzik, and A. Sanpera. Qnd for spinor gases. *Nature Physics*, 4:50, 2008.

[41] I. B. Mekhov and H. Ritsch. Quantum nondemolition measurements and state preparation in quantum gases by light detection. *Phys. Rev. Lett.*, 102:020403, 2009.

[42] G. De. Chiara, O. Romero-Isart, and A. Sanpera. Probing magnetic order in ultracold lattice gases. *arXiv:1007.2591 (cond-mat.quant-gas)*, 2010.

[43] P. Bushev, D. Rotter, A. Wilson, F. Dubin, C. Becher, J. Eschner, R. Blatt, V. Steixner, P. Rabl, and P. Zoller. Feedback cooling of a single trapped ion. *Phys. Rev. Lett.*, 96:043003, 2006.

[44] J. Javanainen. Spectrum of light scattered from a degenerate bose gas. *Phys. Rev. Lett.*, 75:1927, 1995.

[45] T. L. Dao, A. Georges, J. Dalibard, C. Salomon, and I. Carusotto. Measuring the one-particle excitations of ultracold fermionic atoms by stimulated raman spectroscopy. *Phys. Rev. Lett.*, 98(24):240402, 2007.

[46] Tung-Lam Dao, Iacopo Carusotto, and Antoine Georges. Probing quasiparticle states in strongly interacting atomic gases by momentum-resolved raman photoemission spectroscopy. *Phys. Rev. A*, 80(2):023627, 2009.

[47] M. Saba, T. A. Pasquini, C. Sanner, Y. Shin, W. Ketterle, and D. E. Pritchard. Light scattering to determine the relative phase of two bose-einstein condensates. *Science*, 307:1945, 2005.

[48] J. Dupont-Roc C. Cohen-Tannoudji and G. Grynberg. *Photons and Atoms, Introduction to Quantum Electrodynamics*. John Wiley & Sons, Inc., 1989.

[49] W. P. Schleich. *Quantum Optics in Phase Space*. WILEY VCH, Berlin, 1. edition, 2001.

[50] P. .W. Milonni. *The Quantum Vaccum*. Academic Press, Berlin, 1. edition, 1994.

[51] H. A. Bachor and T. C. Ralph. *A Guide to Experiments in Quantum Optics*. WILEY VCH, Berlin, 2. edition, 2004.

BIBLIOGRAPHY

[52] M. Gross and S. Haroche. Superradiance: An essay on the theory of collective spontaneous emission. *Phys. Rep.*, 93:301, 1982.

[53] A. A. Abrikosov, L. P. Gorkov, and I. E. Dzyaloshinski. *Methods of Quantum Field Theory in Statistical Physics*. Dover Publications, Inc., 1. edition, 1975.

[54] M. Lewenstein, L. You, J. Cooper, and K. Burnett. Quantum field theory of atoms interacting with photons: Foundations. *Phys. Rev. A*, 50:2207, 1994.

[55] J. von Neumann. *Mathematische Grundlagen der Quantenmechanik*. Julius Springer, Berlin, 1. edition, 1932.

[56] M. Göppert-Mayer. über elementarakte mit zwei quantensprüngen. *Ann. Phys. (Leipzig)*, 9:273, 1931.

[57] D. A. Bromley and W. Greiner. *Classical Electrodynamics*. Springer, 1. edition, 1998.

[58] W. Heitler. *The Quantum Theory of Radiation*. Dover Publications, Inc., 3 edition.

[59] H. J. Metcalf and P. van der Straten. *Laser Cooling and Trapping*. Springer, 1. edition, 1999.

[60] L. D. Landau and E. M. Lifschitz. *Quantum Mechanics*. New York, Pergamon, 3. edition, 1977.

[61] L. Pitaevskii and S. Stringari. *Bose Einstein Condensation*. Oxford Science Publications, 2003.

[62] O. Dulieu and C. Gabbanini. The formation and interactions of cold and ultracold molecules: new challenges for interdisciplinary physics. *Rep. Prog. Phys.*, 72:086401, 2009.

[63] M. Conbescot, O. Betbeder-Matibet, and F. Dubin. The many-body physics of composite bosons. *Phys. Rep.*, 463:215, 2008.

[64] T. F. Gallagher. Rydberg atoms. *Rep.Prog.Phys.*, 51:143, 1988.

[65] P. Verkerk, B. Lounis, C. Salomon, C. Cohen-Tannoudji, J. Y. Courtois, and G. Grynberg. Dynamics and spatial order of cold cesium atoms in a periodic optical potential. *Phys. Rev. Lett.*, 68(26):3861–3864, 1992.

[66] P. S. Jessen, C. Gerz, P. D. Lett, W. D. Phillips, S. L. Rolston, R. J. C. Spreeuw, and C. I. Westbrook. Observation of quantized motion of rb atoms in an optical field. *Phys. Rev. Lett.*, 69(1):49–52, 1992.

[67] M. P. A. Fisher, P. B. Weichman, G. Grinstein, and D. S. Fisher. Boson localization and the superfluid-insulator transition. *Phys. Rev. B*, 40(1):546–570, 1989.

[68] S. Sachdev. *Quantum Phase Transitions*. Cambridge University Press, 1. edition, 1999.

[69] C. Cohen-Tannoudji, J. Dupont-Roc, and G. Grynberg. *Atom-Photon Interactions, Basic Processes and Applications*. Wiley eds., 2004.

[70] O. Mandel, M. Greiner, A. Widera, T. Rom, T. W. Hänsch, and I. Bloch. Coherent transport of neutral atoms in spin-dependent optical lattice potentials. *Phys. Rev. Lett.*, 91:010407, 2003.

[71] N. W. Ashcroft and N. D. Mermin. *Solid State Physics*. Saunders College Publishing, Philadelphia, 1. edition, 1976.

[72] W. Kohn. Analytical prperties of wannier functions. *Phys. Rev.*, 115:809, 1959.

[73] J. Larson, S. Fernandez-Vidal, G. Morigi, and M. Lewenstein. Quantum stability of mott-insulator states of ultracold atoms in optical resonators. *New J. Phys.*, 10:045002, 2008.

[74] T. Stöferle, H. Moritz, C. Schori, M. Köhl, and T. Esslinger. Transition from a strongly interacting 1d superfluid to a mott insulator. *Phys. Rev. Lett.*, 92:130403, 2004.

[75] S. Wessel, F. Alet, M. Troyer, and G. Batrouni. *Quantum Monte Carlo Simulation of Confined Bosonic Atoms in Optical Lattices in ADVANCES IN SOLID STATE PHYSICS*. Springer, 2004.

[76] R. P. Feynman. *Statistical Mechanics*. W. A. Benjamin, Inc., Massachusetts, 1. edition, 1972.

[77] A. M. Rey. *Ultracold bosonic atoms in optical lattices*. PhD thesis, University of Maryland, 2004.

[78] Stephen B. Haley and Paul Erdös. Standard-basis operator method in the green's-function technique of many-body systems with an application to ferromagnetism. *Phys. Rev. B*, 5(3):1106–1119, 1972.

[79] Y. Ohashi, M. Kitaura, and H. Matsumoto. Itinerant-localized dual character of a strongly correlated superfluid bose gas in an optical lattice. *Phys. Rev. A*, 73:033617, 2006.

[80] A. L. Fetter and J. D. Walecka. *Quantum Theory of Many-Particle Systems*. McGraw-Hill, San Francisco, 1971.

[81] C. Menotti and N. Trivedi. Spectral weight redistribution in strongly correlated bosons in optical lattices. *Phys. Rev. B*, 77(23):235120, 2008.

[82] G. Grynberg and C. Robilliard. Cold atoms in dissipative optical lattices. *Phys. Rep.*, 355:335, 2001.

[83] I. H. Deutsch, R. J. C. Spreeuw, S. L. Rolston, and W. D. Phillips. Photonic bandgaps in optical lattices. *Phys. Rev. A*, 52:1394, 1995.

[84] Y. D. Chong, D. E. Pritchard, and M. Soljacic. Quantum theory of a resonant photonic crystal. *Phys. Rev. B*, 75:235124, 2007.

[85] D. V. van Coevorden, R. Sprik, A. Tip, and A. Lagendijk. Photonic band structure of atomic lattices. *Phys. Rev. Lett.*, 77:2412, 1996.

[86] P. Lambropoulos, G. M. Nikolopoulos, T. R. Nielsen, and S. Bay. Fundamental quantum optics in structured reservoirs. *Rep. Prog. Phys.*, 63:455, 2000.

[87] M. Artoni, G. La Rocca, and F. Bassani. Resonantly absorbing one-dimensional photonic crystals. *Phys. Rev. E*, 72:046604, 2005.

[88] F. Bariani and I. Carusotto. Light propagation in atomic mott insulators. *Eur. Opt. Soc.*, 3:08005, 2008.

[89] N. Bar-Gill, R. Pugatch, E. Rowen, N. Katz, and N. Davidson. Ultracold atoms in incommensurable 1d optical lattices - an interacting aubry-andré model. *preprint arXiv:cond-mat/0603513*, 2006.

[90] G. Roati, C. D'Errico, L. Fallani, M. Fattori, C. Fort, M. Zaccanti, G. Modugno, M. Modugno, and M. Inguscio. Anderson localization of a non-interacting bose-einstein condensate. *Nature*, 453:895, 2008.

[91] P. Barmettler, A. M. Rey, E. Demler, M. D. Lukin, I. Bloch, and V. Gritsev. Quantum many-body dynamics of coupled double-well superlattices. *Phys. Rev. A*, 78:012330, 2008.

[92] S. Fölling, S. Trotzky, P. Cheinet, M. Feld, R. Saers, A. Widera, T. Müller, and I. Bloch. Direct observation of second order atom tunnelling. *Nature*, 448:1029, 2007.

[93] J. E. Lye, L. Fallani, C. Fort, V. Guarrera, M. Modugno, D. S. Wiersma, and M. Inguscio. Direct observation of second order atom tunnelling. *Phys. Rev. A*, 75:061603(R), 2007.

[94] S. Stenholm. The semiclassical theory of laser cooling. *Rev. Mod. Phys.*, 58:699, 1986.

[95] H. J. Kimble. *Cavity Quantum Electrodynamics*. Academic, New York, 1. edition, 1994.

[96] D. E. Chang, V. Gritsev, G. Morigi, V. Vuletic, M. D. Lukin, and E. A. Demler. Crystallization of strongly interacting photons in a nonlinear optical fibre. *Nature physics*, 4:884, 2008.

[97] C. A. Christensen, S. Will, M. Saba, G. B. Jo, Y. I. Shin, W. Ketterle, and D. Pritchard. Trapping of ultracold atoms in a hollow-core photonic crystal fiber. *Phys. Rev. A*, 78:033429, 2008.

[98] T. Holstein and H. Primakoff. Field dependence of the intrinsic domain magnetization of a ferromagnet. *Phys. Rev.*, 58:1098, 1940.

[99] P. Stehle. Atomic radiation in a cavity. *Phys. Rev. A*, 2(1):102–106, 1970.

[100] S. Slama, C. von Cube, A. Ludewig, M. Kohler, C. Zimmermann, and Ph. W. Courteille. Dimensional crossover in bragg scattering from an optical lattice. *Phys. Rev. A*, 72:031402(R), 2005.

[101] S. Slama, C. von Cube, B. Deh, A. Ludewig, C. Zimmermann, and Ph. W. Courteille. Phase-sensitive detection of bragg scattering at 1d optical lattices. *Phys. Rev. Lett.*, 94:193901, 2005.

[102] S. Slama, C. von Cube, M. Kohler, C. Zimmermann, and Ph. W. Courteille. Multiple reflections and diffuse scattering in bragg scattering at optical lattices. *Phys. Rev. A.*, 73:023424, 2006.

[103] D. F. Walls and G. J. Milburn. *Quantum Optics*. Springer, Berlin, 1. edition, 1994.

[104] P. R. Rice and R. J. Brecha. Cavity induced transparency. *Opt. Comm.*, 126:230, 1996.

[105] D. Kruse, M. Ruder, J. Benhelm, C. von Cube, C. Zimmermann, P. W. Courteille, T. Elsässer, B. Nagorny, and A. Hemmerich. Cold atoms in a high-q ring cavity. *Phys. Rev. A*, 67:051802(R), 2003.

[106] S. Fernandez-Vidal, S. Zippilli, and G. Morigi. Nonlinear optics with two trapped atoms. *Phys. Rev. A*, 76:053829, 2007.

[107] W. M. Itano, J. J. Bollinger, J. N. Tan, B. Jelenkovic, X. P. Huang, and D. J. Wineland. Bragg diffraction from crystallized ion plasmas. *Science*, 279:686, 1998.

[108] M. Weidemüller, A. Hemmerich, A. Görlitz, T. Esslinger, and T. W. Hänsch. Bragg diffraction in an atomic lattice bound by light. *Phys. Rev. Lett.*, 75:4583, 1995.

[109] G. Birkl, M. Gatzke, I. H. Deutsch, S. L. Rolston, and W. D. Phillips. Bragg scattering from atoms in optical lattices. *Phys. Rev. Lett.*, 75:2823, 1995.

[110] L. Guidoni, C. Triché, P. Verkerk, and G. Grynberg. Quasiperiodic optical lattices. *Phys. Rev. Lett.*, 79:3363, 1997.

[111] S. Pirandola, S. Mancini, D. Vitali, and P. Tombesi. Continuous-variable entanglement and quantum-state teleportation between optical and macroscopic vibrational modes through radiation pressure. *Phys. Rev. A*, 68:062317, 2003.

[112] J. I. Cirac, R. Blatt, A. S. Parkins, and P. Zoller. Spectrum of resonance flourescence from a single trapped ion. *Phys. Rev. A*, 48:2169, 1993.

[113] S. Mancini, D. Vitali, and P. Tombesi. Scheme for teleportation of quantum states onto a mechanical resonator. *Phys. Rev. Lett.*, 90:137901, 2003.

[114] G. Morigi, J. Eschner, S. Mancini, and D. Vitali. Entangled light pulses from single cold atoms. *Phys. Rev. Lett.*, 96:023601, 2006.

[115] T. D. Kühner, S. R. White, and H. Monien. One-dimensional bose hubbard model with nearest neighbour interaction. *Phys. Rev. B*, 61:12474, 2000.

[116] K. Sengupta and N. Dupuis. Mott-insulator-to-superfluid transition in the bose-hubbard model: A strong-coupling approach. *Phys. Rev. A*, 71:033629, 2005.

[117] S. Konabe, T. Nikuni, and M. Nakamura. Laser probing of the single-particle energy gap of a bose gas in an optical lattice in the mott-insulator phase. *Phys. Rev. A*, 73:033621, 2006.

[118] S. D. Huber, E. Altman, H. P. Büchler, and G. Blatter. Dynamical properties of ultracold bosons in an optical lattice. *Phys. Rev. B*, 75:085106, 2007.

[119] C. Menotti and N. Trivedi. Spectral weight redistribution in strongly correlated bosons in optical lattices. *Phys. Rev. B*, 77:235120, 2008.

[120] K. Huang. *Quantum Field Theory*. WILEY VCH, 2. edition, 2010.

[121] D. Pines and P. Nozieres. *The theory of quantum liquids*. Benjamin, New York, 1. edition, 1966.

[122] C. J. Pethick and H. Smith. *Bose Einstein Condensation in Dilute Gases*. Cambridge University Press, 2008.

[123] F. Englert and R. Brout. Broken symmetry and the mass of gauge vector mesons. *Phys. Rev. Lett.*, 13(9):321–323, 1964.

[124] P. W. Higgs. Broken symmetries and the masses of gauge bosons. *Phys. Rev. Lett.*, 13(16):508–509, 1964.

[125] D. L. Luxat and A. Griffin. Coherent tunneling of atoms from bose-condensed gases at finite temperatures. *Phys. Rev. A*, 65:043618, 2002.

[126] S. Choi, Y. Japha, and K. Burnett. Adiabatic output coupling of a bose gas at finite temperatures. *Phys. Rev. A*, 61(6):063606, 2000.

[127] R. P. Feynman and A. R. Hibbs. *Quantum Mechanics and Path Integrals*. New York: McGraw-Hill, 1965.

[128] B. G. Englert. Fringe visibility and which-way information: An inequality. *Phys. Rev. Lett.*, 77:2154, 1996.

[129] Marlan O. Scully and Kai Drühl. Quantum eraser: A proposed photon correlation experiment concerning observation and "delayed choice" in quantum mechanics. *Phys. Rev. A*, 25(4):2208, 1982.

[130] Y. Shin, G. B. Jo, M. Saba, T. A. Pasquini, W. Ketterle, and D. E. Pritchard. Optical weak link between two spacially separated bose-einstein condensates. *Phys. Rev. Lett.*, 95:170402, 2005.

[131] M. Saba, T. A. Pasquini, C. Sanner, Y. Shin, W. Ketterle, and D. E. Pritchard. Light scattering to determine the relative phase of two bose-einstein condensates, "suplementary information". *Science*, 307:1945, 2005.

[132] M. O. Scully and M. S. Zubairy. *Quantum Optics*. Cambridge University Press, 1. edition, 1997.

[133] P. C. Hohenberg. Existence of long-range order in one and two dimensions. *Phys. Rev.*, 158(2):383–386, 1967.

[134] W. Ketterle and N. J. van Druten. Bose-einstein condensation of a finite number of particles trapped in one or three dimensions. *Phys. Rev. A*, 54(1):656, 1996.

[135] V. B. Braginsky and F. Y. Khalili. *Quantum Measurement*. Cambridge University Press, 1. edition, 1992.

[136] M. D. Levenson, R. M. Shelby, M. Reid, and D. F. Walls. Quantum nondemolition detection of optical quadrature amplitudes. *Phys. Rev. Lett.*, 57(20):2473–2476, 1986.

[137] A. Kuzmich, L. Mandel, J. Janis, Y. E. Young, R. Ejnisman, and N. P. Bigelow. Quantum nondemolition measurements of collective atomic spin. *Phys. Rev. A*, 60(3):2346–2350, 1999.

[138] F. Brennecke, T. Donner, S. Ritter, T. Bourdel, M. Köhl, and T. Esslinger. Cavity qed with a bose-einstein condensate. *Nature*, 450:268, 2007.

[139] Y. Colombe, T. Steinmetz, G. Dubois, F. Linke, D. Hunger, and J. Reichel. Strong atom-field coupling for bose-einstein condensates in an optical cavity on a chip. *Nature*, 450:272, 2007.

[140] K. Baumann, C. Guerlin, F. Brennecke, and T. Esslinger. Dicke quantum phase transition with a superfluid gas in an optical cavity. *Nature*, 464:1301, 2010.

[141] S. Fernández-Vidal, G. De Chiara, J. Larson, and G. Morigi. Quantum ground state of self-organized atomic crystals in optical resonators. *Phys. Rev. A*, 81(4):043407, 2010.

[142] J. J. Sakurai. *Modern Quantum Mechanics*. Addison-Wesley, New York, 1. edition, 1994.

[143] A. Griffin. Conserving and gapless approximations for an inhomogeneous bose gas at finite temperatures. *Phys. Rev. B*, 53:9314, 1996.

[144] S. Giorgini, L. Pitaevskii, and S. Stringari. Thermodynamics of trapped bose-condensed gas. *Rev. Mod. Phys.*, 109:309, 1997.

[145] M. Abramowitz and I. A. Stegun. *Handbook of mathematical functions*. Dover Publications Inc., New York, 1968.

I want morebooks!

Buy your books fast and straightforward online - at one of world's fastest growing online book stores! Environmentally sound due to Print-on-Demand technologies.

Buy your books online at
www.morebooks.shop

Kaufen Sie Ihre Bücher schnell und unkompliziert online – auf einer der am schnellsten wachsenden Buchhandelsplattformen weltweit! Dank Print-On-Demand umwelt- und ressourcenschonend produziert.

Bücher schneller online kaufen
www.morebooks.shop

KS OmniScriptum Publishing
Brivibas gatve 197
LV-1039 Riga, Latvia
Telefax: +371 686 204 55

info@omniscriptum.com
www.omniscriptum.com

Printed by Books on Demand GmbH, Norderstedt / Germany